Adam Hart-Davis

SCIENTIFIC EYE

Illustrated by
Jane Cope

Designed by
Geoffrey Wadsley

Unwin Hyman Limited

Contents

Acknowledgements

The author would like to acknowledge that the ideas in this book came from a variety of sources, including David Attenborough's *Life on Earth*, Michael Faraday's *The Chemical History of a Candle*, Desmond Morris's *The Naked Ape*, Bram Stoker's *Dracula*, and the BBCTV Horizon programme *The Life that Lives on Man*. Many people made positive suggestions about what I should do with the text. Those who were particularly helpful in shaping it include David Edwards, Jo Heys, David Jones, Christopher Kington, Robert Moss, Antonia Murphy, Margaret Sands, David Waddington, John Walker, Martin Ward and Simon Welfare. Thank you all.

The Photographs come from:
Seaphot Limited, **11**; Met Office, **28**, **31**; Dr Y Furukawa **31** Dr Charles Knight **31**; Natural History Photographic Agency, **32**; Science Photo Library, **46**; Zoological Society of London, **46**; Ministry of Agriculture, Fisheries & Food, **46**, **58**; Biofotos, **54**; Natural History Museum **54**; Royal Veterinary College; **58**. All other photographs from Adam Hart-Davis.

The Scientific Eye television series is a Yorkshire Television production. Executive Producer: Chris Jelley. Producer: Adam Hart-Davis. Director: Michael Cocker.

Published by
UNWIN HYMAN LIMITED
15/17 Broadwick Street
London W1V 1FP

Reprinted 1986, 1987

© Adam Hart-Davis, 1985

ISBN 0 7135 2493 6 (limp)
0 7135 2584 3 (cased)

Typeset by August Filmsetting, Haydock, St Helens
Reproduction by Positive Colour Maldon, Essex
Printed in Great Britain by Scotprint Ltd., Musselburgh.

Why bother with science?

Best bubble-gum bubbles better than second-best bubble-gum. Why do you think that should be? Perhaps there are bubble-gum scientists studying the bubbles using all the latest instruments. When they understand more clearly how bubble-gum works, they may be able to make it bubble better.

Scientists are people who try to understand how things work.

Science is what they use to try and understand.

The scientific method

Suppose that outside a school there were suddenly several accidents in which children were knocked down by cars. How could they be stopped? There could be many suggestions put forward. How can you tell which is right?

This does not sound like a scientific problem. Even so, you can try and solve it by using the **scientific method**. Science is not just about experiments. Science is a way of looking at the world. Science is about rubber bands, and cats, and custard. People who think like scientists have a particular way of looking at the world, and at its problems.

The **first** part of the scientific method is to gather information. **Second**, sort it out, or *analyse* it. **Third**, guess what is going on. **Fourth**, test your guess with one or more experiments.

What was the information about the accidents outside the school? It turned out that they all happened at the same time of day, during morning break. And there was a new ice-cream van which had started coming up from the town to park opposite the school.

Now you are in a position to guess what was happening. Children were running across the road to get to the ice-cream van. The solution was simple. The driver was asked to park on the same side of the road as the school.

The council tested this solution by watching for further accidents. No more were reported. The solution was successful.

Remember the scientific method: observe, analyse, make a theory, test it by experiment. Or if you like **Look, Think, Guess, Try.**

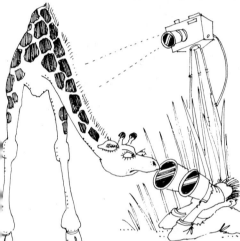

What do scientists do?

Some people think that a scientist is a person who wears a white coat and talks about experiments. Well, some scientists do. Others climb mountains, go to Africa to look for strange animals, do delicate brain surgery, or work out ways of producing more food.

Two black cats want to get through a door. The door is slightly open, but it opens towards them; so when they push they can't get through. Sam, the dim cat, sits there and mews, in the hope that someone will come and open the door. But Jack does not give up so easily. He pushes his nose through, and then his paw, trying to reach outside. And in due course he finds that if he puts his paw through and pulls, then he can open the door.

You could not call Jack a scientist. His scientific method is mainly trial and error. But he does solve problems, and he does try to understand his environment. That is what science is all about.

The point of reading this book is to find out about thinking scientifically. By thinking scientifically you can better understand how the whole world works. Reading this book won't turn you into a scientist, but it will help you to look at the world and its problems with a scientific eye.

Averages, pies, and bars

Suppose you have chosen a problem to study. The first stage of the scientific process is watching. You must gather information before you can do anything else. And what you usually gather is a whole load of numbers. What scientists often need is a way to make sense of these numbers.

Averages

Suppose you asked all the girls in your class how long it takes to walk to school in the morning. Their answers, in minutes, are 3, 8, 24, 7, 14, 30, 12, 9, 10, and 33.

Now you can guess that they don't all walk at the same speed. And some will have further to walk than others. But you can get a general idea of how long it takes a girl to walk to school from the **average** time. You get the average by adding all the answers together, and then dividing by the number of answers.

$$\text{Average} = \frac{\text{Sum of all answers added together}}{\text{Number of answers}}$$

In this case there are ten girls, and the average time they take to walk to school is

$$\frac{3+8+24+7+14+30+12+9+10+33}{10} = \frac{150}{10} \text{ or 15 minutes}$$

Notice that none of the girls takes exactly 15 minutes; the average number does not have to be one of the answers.

If you drink one glass of milk every day, and your friend drinks three, what is the average number of glasses of milk drunk every day?

Total number of glasses of milk = 4
Total number of children = 2
Average = $\frac{4}{2}$ = 2 glasses of milk every day

Pies

Pie charts are a splendid way of displaying information. Suppose you went round the whole class and asked everyone 'What's your favourite flavour of crisps?' The answers are 15 cheese and onion, 8 ready salted, 6 smoky bacon, and 1 salt and vinegar.

cheese & onion — 15
ready salted — 8
smoky bacon — 6
salt & vinegar — 1
———
30

You now want to show what fraction of the class likes each flavour. Draw a **pie chart**. The pie stands for the whole class. Exactly half the children (15 out of 30) like cheese and onion. So colour half the pie for cheese and onion. Just over half the rest like ready salted; colour them a thick slice. One fifth of the class like smoky bacon; colour them a thin slice. And colour a tiny piece for the one who likes salt and vinegar. Colour the small slices the darkest, so you can see them.

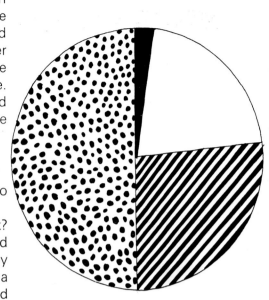

Bars

Bar charts are better than pies when you have numbers to deal with.

How many apples do the boys in your class eat in a week? First make a table with 'number of apples' across the top and 'number of boys' up the side. Go round and ask each boy how many apples he eats, and fill in the table. Then make a bar chart just like the table, but using coloured bars instead of ticks. From the chart you can see at once which is the most common number, and how many boys eat very few apples.

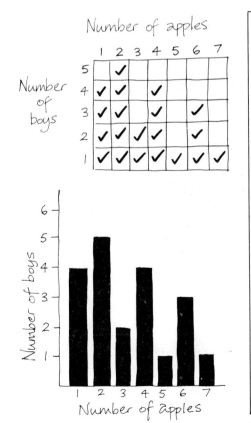

Number of apples

Number of boys

Number of apples

Questions

1 The Smiths at 1 North Road have 3 children. The Joneses at No 2 have 2 children. The family at No 3 have 2 boys, 2 girls, and a dog. What is the average number of children in the three houses?

   The Smiths move out and are replaced by a couple with 5 cats but no children. What is the average number of children in the three houses now?

2 Go round your own class and find out which are the favourite flavours of crisps. Draw a pie chart of the results. Explain how it is different from the one above.

3 Find out the ages, in years and months, of the children in your class. Make a bar chart of the ages. Suggest a way of showing both boys and girls separately on the same bar chart.

   What is the average age of the boys?
   What is the average age of the girls?
   Are these the same as the most common ages? If not, can you suggest reasons why they might be different?

1. Mass meeting

If you turn into a ghost, **mass** is what you'll miss the most.

Harry Martindale was fixing pipes in a cellar in York. Suddenly, he says, he heard the sudden blast of a trumpet. He looked down, and saw a Roman soldier walk out of the wall.

Terrified, he fell off his ladder, and hid at the back of the cellar. As he watched, a large shaggy horse came through the wall, with another soldier riding it. Then a whole column of Roman soldiers, marching in step. Their helmets shone. Tufts of beard poked through the metal at the chin. Their waistcoats were strips of leather. They wore faded green skirts, and short swords.

The strangest thing was that he could not see their legs below the knees. Later he found that the old Roman road ran 50 cm below the earth and stones of the cellar floor.

Harry was scared stiff that the soldiers would look to their right, see him, and cut him to pieces. But he need not have worried. Like all good ghosts, those Roman soldiers had no mass.

What is mass?

Mass is the amount of stuff in an object. A massive object has a lot of mass in it.

We measure mass in tonnes, kilograms (kg), grams (g), and milligrams (mg).

1 tonne = 1000 kg   1 kg = 1000 g   1 g = 1000 mg

You can't push a pencil through your desk, because the stuff of the pencil bashes into the stuff of the desk. But if the pencil had no mass it would have no stuff, and it could go through the desk. Harry's Roman soldiers were able to march through the wall because they had no mass. And so they could not have hurt him with their swords. The swords would have gone through him, but he would not have felt a thing!

Why mass matters

What can hurt you is to run into someone else at full speed. If you dash along the corridor at school, and go straight into someone else in your class, who would hurt you most? The smallest, or the biggest in the class?

You would knock the smallest person flying. But if you smacked into the biggest you might end up flat as a pancake. Why? Because the biggest in the class is the most massive. In a collision, the more the mass the bigger the bang.

Questions

1 Britain's biggest cabbage (grown in 1977 by P. G. Barton of Cleckheaton) had a mass of 51.8 kg. Is this more massive than you? Write down your mass in kg and then in tonnes.
2 Could a ghost play tennis? Give reasons for your answer.
3 In the supermarket, red potatoes are 20p per pound (lb). White potatoes are 40p per kilo (kg). Which are cheaper per unit of mass? (*Hint:* 1 kg = 2.2 lb. How many kg of each colour do you get for 20p?)
4 Cracking nuts is easier with a big hammer than with a little hammer. Why should this be so?

2. Weight and see

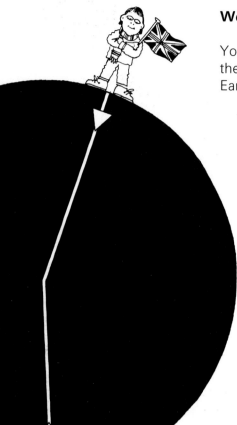

Weight is different from mass.

Your mass is the amount of stuff in your body. The mass of the Earth is the amount of stuff in the Earth. The mass of the Earth pulls your mass down. What you feel is **weight**.

Your weight is the pull of the Earth's mass on your mass.

Wherever you are on Earth, your weight is a downwards pull, never upwards or sideways. Down means towards the middle of the Earth.

Scientists measure weight in newtons. On Earth, each kilogram of mass has a weight of 9.8 newtons, which is almost 10 newtons. So if your mass is 30 kg you weigh nearly 300 newtons.

How to lose weight

Your mass is the same wherever you are, but your weight may be different in different places. Climb a high mountain and you will weigh a little less. Why? Because you are further from the centre of the Earth. Your mass is pulled down less hard.

On the Moon you would weigh only about one sixth of what you weigh on Earth.

Feeling 'weightless'

Astronauts feel 'weightless' as they orbit round the Earth. You can get this 'weightless' feeling too.

Stand on a table. In each hand hold one end of a 1 m piece of string with a roll of sellotape hung over the middle. Keep your arms straight. Start with the string stretched tight. Then bring your hands together. The string will sag as the weight of the sellotape pulls it down.

Now do it again, but this time jump off the table just before you clap your hands together. Jump upwards off the table so that you are in the air as long as possible. Clap your hands together while you are falling. The string will not sag. You will catch the sellotape. While you are falling, both you and the sellotape seem to have no weight.

Gravity

Any two lumps of stuff in the Universe are pulled towards one another. This pull is called **gravity.** The Earth has a big mass, and so its gravity is usually the biggest pull on your body. This pull of gravity is what gives you weight. On the Moon, gravity is six times smaller than it is on Earth.

Not much gravity up here mate!

MOON

Questions

1 What is the difference between a ghost and an astronaut?
2 The weight of a thing on Mars is about half its weight on Earth. How much (in newtons) do you weigh on Earth? How much would you weigh on Mars? Why should you weigh less there? Do you think Martians would be good at the high jump? Why?
3 Find out the weight (in newtons) of each member of the class. Make a bar chart, colouring one square for each person. Which is the most common weight? What is the average weight of the class? How could you find out whether the boys are on average heavier than the girls?
4 Copy and complete this wordsquare. Write down the meaning of the word in the box down the middle.

1 The unit of weight.
2 Something with lots of weight feels this.
3 The pull of this is what gives you weight.
4 One of these has a weight of 9.8 newtons . . .
5 . . . on this planet.
6 This person sometimes feels weightless.

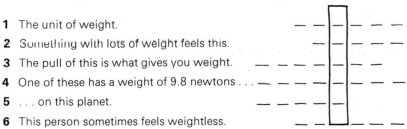

3. May the force be with you

If you want to move a mountain, or a molehill, or a horse,
When the ketchup bottle's blocked with solidified red sauce,
If you want to get things moving, then what you need is force.

Force makes things move

Forces move things. Nothing can *begin* to move unless a force starts it. Forces may push, pull, tug, heave, squeeze, stretch, twist, or press, but what they all do is move things.

Weight is a force. Your mass is pulled down by gravity. The result of that pull is the force we call weight. Does this force make you move? Yes. Step off a chair, and the force will pull you quickly to the floor. Sit at the top of a slide in the playground, and the force will pull you down.

Sit on a springy bench, and the bench will bend under your weight. The force of your weight pulls you down until the bench has bent enough to hold you. The more mass you have the bigger your weight, and the more you will bend the bench.

When does the bench stop bending? When the forces balance. The bench is springy. It pushes upwards on you. The more it bends the harder it pushes. When the push of the bench is exactly equal to your weight then the bench will stop bending.

WEIGHT

FORCE

SQUEEZE

PRESS

PUSH

BEND

FLATTEN

How do we measure force?

Forces are measured in newtons. If you hang a 1 kg mass on a piece of string, it will pull on the string with a force of 9.8 newtons, or nearly 10 newtons.

Weight is a force. If your mass is 40 kg, then your weight is nearly 400 newtons.

Questions

1 Make a list of ten things forces can do.
2 A pulling force in a string or rope is called **tension**. A 2 kg bunch of bananas hangs on a piece of string. What is the tension in the string?
3 Copy and complete the wordsquare below.

HEAVE

PULL

TWIST

SQUASH

Remember: forces move things.
If something isn't moving,
then the forces on it are exactly
balanced.

Britain's biggest weighed more than 500 newtons

1 What you need to move things.
2 The force of this gives you weight.
3 When you sit on a bench your weight will make it do this . . .
4 . . . until the forces do this.
5 If you weigh 500 newtons, you have 50 kg of this.
6 The force that pulls you down.
7 Force pulling in a string.

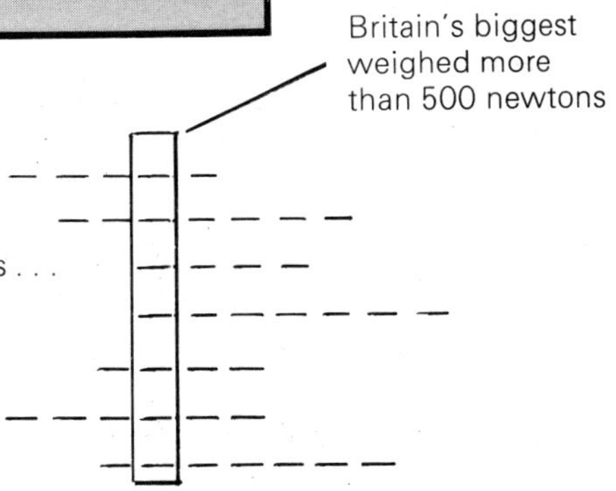

4. Science friction

Why do PE shoes have rubber soles? To help stop you slipping on the floor. Rubber doesn't slip easily on a wooden floor, because of **friction**. Friction helps to stop slipping.

When one thing slides over another – like a foot over the floor – friction is a force that tries to stop it. The size of this force depends on the floor, and on the shoes. Leather is more slippery than rubber, unless the floor is wet.

The force of friction is greater for things that are heavier. You can see this with a pair of walking legs.

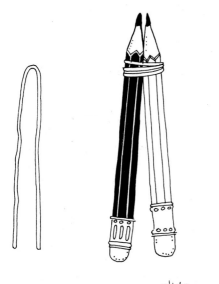

Walking legs

Make a pair of legs by bending a large paperclip into a narrow U-shape. Or take two pencils with rubbers on the ends for feet. Use a rubber band to tie the pencils together, near their points, to make a narrow V-shape.

Hold a ruler on its edge on a desk or table. Stand the legs astride the ruler. They should lean on it at an angle, with both feet on the desk.

Lift the ruler up 3 or 4 mm, and then let it down again. Keep doing this, and the legs will walk along the ruler. They walk better on a rough surface. Try this book if the desk is too slippery.

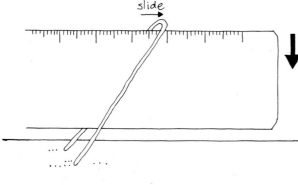

Ruler going down. Weight taken by feet. Friction high at feet, but low at crutch. So crutch slides, but feet do not.

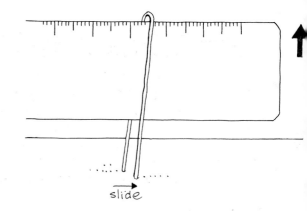

Ruler going up. Weight taken at top of legs. Friction high here. Little weight on feet; so friction low, and they slide easily.

Making more friction

Some new bottle tops are screwed on much too tight. Often you can't get into the Coke, or the jam jar. The top just slips round in your hand. What you need is *more friction*.

Some people use a tea towel, but old rubber kitchen gloves are better. Cut a finger off the glove to slip over a bottle top. Slip a wrist over a jar lid. The rubber glove will make more friction. You will get a better grip of the lid, and you should be able to unscrew it.

Making less friction

Old bicycles squeak and stick. This means too much friction. Put a drop of oil in the right place and the bike should run smoothly again. Sticking is caused by friction. Oil reduces friction. Oil is a **lubricant**.

Wet soap is also a lubricant. Have you ever dropped the soap in the shower or in the bath? It is hard to pick up, because it is so slippery. If you tread on a piece of wet soap you will probably fall flat on your face!

Banana skins are even more slippery than oil or wet soap. The outside of a banana skin is like rubber – not slippery at all. But the inside has a thick layer of sugary paste. This paste is very slippery indeed. If you tread on it you will almost certainly slip.

What's even worse is to tread on the outside. Then the inside of the skin will slide on the floor. If you tread on the inside you can skid only the length of the skin, perhaps 20 cm. But if you tread on a banana skin slippery side down you could skid for miles . . .

Questions

1 Suggest reasons why the walking legs work better on a rough surface.
2 When you want to walk forwards, you push back on the ground with your foot.
  a Why doesn't your foot slide backwards?
  b Which of these surfaces would give you enough friction to walk on easily: wet soap, sandpaper, concrete, banana skins?
3 Write a short letter to a shoe maker to explain why trainers need to have rough soles.

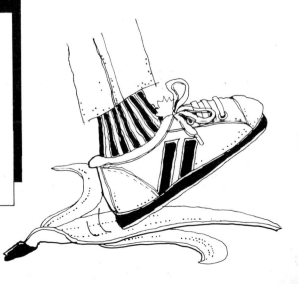

5. The long and the short and the tall

In 1791 Napoleon was only a junior officer in the French army. That year three French scientists said that everyone should measure lengths and distances in metres. Not gasometers, nor thermometers, but plain metres. 'How long is a metre?' said everyone. Their answer was that ten million metres would be the distance from the equator to the north pole. So one metre was to be one ten millionth of that distance.

The geography experts took eight years to find out how far this was! Even then they didn't get it quite right – but now the metre is the basic unit of length. Just as with grams, there are names for other useful units of length. Copy this table into your book.

Metric system of length and distance
1 kilometre (km) = 1000 metres (m)
1 metre (m) = 100 centimetres (cm)
1 metre (m) = 1000 millimetres (mm)

Remember
kilo = 1000 ×
milli = 1/1000 ×

Look at your ruler. You will see centimetres marked on one side, and divided into millimetres. Write down and complete this sentence, and add it to the table in your book:

1 centimetre (cm) = ?? millimetres (mm)

The reason for having all these names is that they make life easier. The height of your desk is probably just less than a metre, perhaps 0.94 m. We know how big a centimetre is, and it's more convenient to say the height is 94 cm.

In days of old, when men were bold, and metres weren't invented,
With cubits, spans, and feet and hands, they had to be contented.

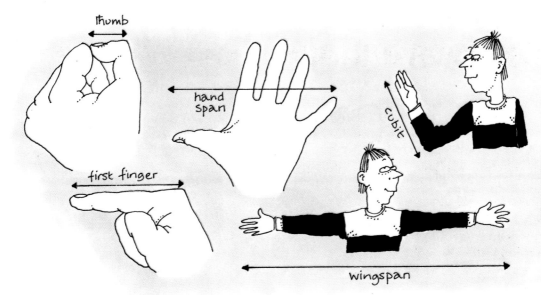

How big are you?

Use your ruler to measure yourself. Write down a list of body measurements, and fill in the lengths in centimetres.

Before people had rulers or metres these were the sort of units they used. They are still useful. If you want to measure the length of a table or the width of a window you can use your body to measure it.

Roman soldiers were measuring distances hundreds of years before Napoleon was born. They used to count their steps. They put a post in the ground every thousand double paces (left . . . right . . .). Latin for 1000 is *mille*, and 1000 double paces came to be called a mile.

Questions

1 a How many mm in 6 m, 7.31 m, $\frac{1}{4}$ m?
  b How many metres in 2300 mm, 4321 mm, 250 cm?
  c How many cm in 2 m, 25 mm, 0.42 m, 2 m 764 mm?

2 A tape measure is a floppy ruler. Use one to measure round the widest part of your head. If you have no tape measure, how can you do it with a piece of string and a ruler?

3 Make a bar chart of the distances round all the heads in the class. Colour boys in one colour, girls in another. Who have the bigger heads? What is the average size?

4 Using only your body, measure the following things:
  a the length of the classroom
  b the height of the windows
  c the length of your pencil
  d the length of the playground

5 Go to the library, and find out how far it is, in kilometres, to
  a the nearest post office
  b Stonehenge
  c the sea side
  d New York

6. No square feet

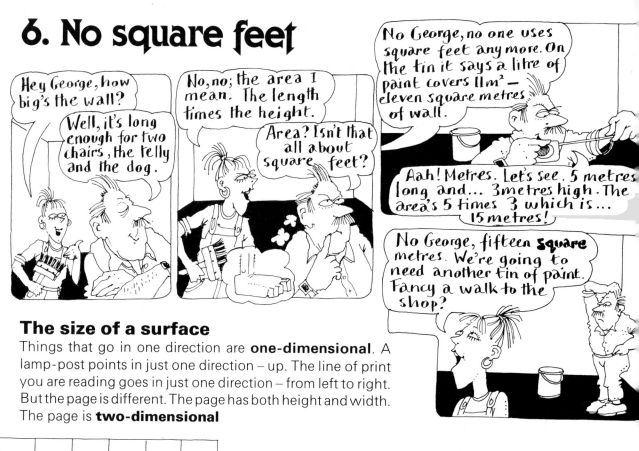

The size of a surface

Things that go in one direction are **one-dimensional**. A lamp-post points in just one direction – up. The line of print you are reading goes in just one direction – from left to right. But the page is different. The page has both height and width. The page is **two-dimensional**

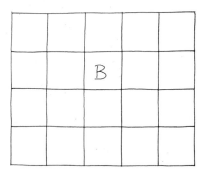

Shape **A** on the left has three rows of squares and six columns. Each square is 1 cm across and 1 cm down; it is one square centimetre (1 cm²). How big is the shape? It is 6 cm long and 3 cm high. The area of the shape is the amount of space it takes up – 18 squares, which is 18 square centimetres (18 cm²).

> *Remember*   m² = square metres (also written sq.m.)
>   cm² = square centimetres (also written sq.cm.)

Shape **B** on the right has four rows of squares, in five columns. Write down the answers to these questions:
a How many squares are there in this shape? **b** What is the area of the shape? **c** Which of the two shapes is bigger?

> Even though shape **A** is longer (6 cm) than shape **B** (5 cm), shape **B** has a larger area. **B** is the bigger shape

Units of area

You don't have to count the squares. It's usually quicker to measure the length and the width, and multiply them together. To find the area of this page you could divide it into 1 cm squares, and count them. But try it the other way. Measure the width of the page to the nearest cm. Then measure the height. Write them both down and multiply them together. You should get between 400 cm² and 500 cm².

If you use centimetres you will get the area in square centimetres. If you use millimetres you will get the answer in square millimetres. What you must never do is mix units. Don't multiply centimetres by millimetres. If you mix units you won't get an area, you'll get a mess!

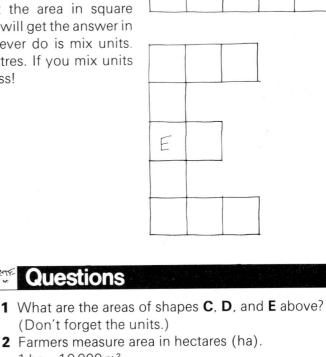

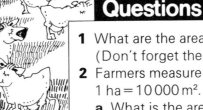

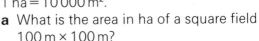

THE TWO HECTARE FIELD

Questions

1 What are the areas of shapes **C**, **D**, and **E** above? (Don't forget the units.)

2 Farmers measure area in hectares (ha). 1 ha = 10 000 m².
   a What is the area in ha of a square field 100 m × 100 m?
   b If each cow needs 20 m², how many cows can fit comfortably into a 2 ha field?

3 Find out the length, width, and height of your classroom. Work out the area (in square metres) of each wall and the ceiling. What is the total area of all these added together? If one litre of paint covers 11 m², how many litres of paint would you need to paint it all?

4 Find out the three *biggest* countries in the world. Write down their names and their areas. Do the same for the three *smallest* countries in *Europe*.

7. Flooding the classroom

How much water would it take to fill your classroom? (DO NOT TRY IT. Your teacher may not approve!) Just suppose that you could find enough chewing gum to block up all the cracks round the windows and the doors. How much water would it take to reach the ceiling?

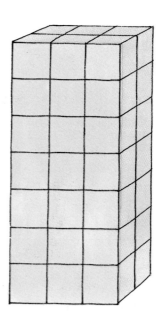

This cube is a solid object with three dimensions – length, width, and thickness. It measures 2 cm long, 2 cm wide, and 2 cm thick. How big is the cube? What is its **volume**?

The cube is made up of eight small cubes. You can see seven, and one is hidden. Each small cube measures 1 cm in each dimension; it is a cubic centimetre. The cube has a volume of eight small cubes, or eight cubic centimetres (8 cm³).

> *Remember*: a **two**-dimensional shape has an **area** that can be measured in **square** centimetres (cm²). A **three**-dimensional object has a **volume** that can be measured in **cubic** centimetres (cm³).

Cubic centimetres (cm³) are sometimes written as cu.cm., or as cc.

Measuring volume

This block is 7 cm high, 3 cm wide, and 2 cm thick.
a Write down the area of the front of the block.
b How many small cubes are there in the whole block?
c What is the volume of the whole block?

---

c Volume of block = 42 cm³.
42 altogether.
b There are 21 small cubes in front, and 21 behind;
a Area of front = 3 cm × 7 cm = 21 cm².

To work out the volume you multiply together the measurements in all three dimensions.

$$\text{Volume} = \text{length} \times \text{height} \times \text{thickness}$$
$$= 3\,\text{cm} \times 7\,\text{cm} \times 2\,\text{cm} = 42\,\text{cm}^3$$

Cubic capacity

Volume is a measure of the space something takes up. It also tells you the **cubic capacity** – how much stuff will fit inside a container. An ordinary milk bottle will hold 1 pint of milk. A plastic bottle may hold 500 millilitres of washing-up liquid.

We measure the volume of solid lumps in cubic centimetres. For runny things we use litres and millilitres. They are all units of volume.

---

1 cubic metre (m³) = 1000 litres (l)
1 litre (l) = 1000 millilitres (ml)
1 millilitre (ml) = 1 cubic centimetre (cm³)

---

1 gallon = 8 pints = about 4.5 l
1 pint = about 568 ml
1 teaspoonful = about 5 ml

Questions

1 A square box is 10 cm long, 10 cm wide, and 10 cm high. What is its volume? How many litres of water will it hold? How many pints of milk could you put in it?

2 Write down the length, width, and height of the classroom, all in metres. Multiply together length × width × height to find the volume of the room (probably between 100 m³ and 200 m³.) How much water would you need to fill this space? Work out the answer in **a** litres and **b** gallons.
  If it takes 20 gallons to fill a bath, how many bathfuls would fit into the classroom?

3 Look at the labels on bottles, tins, and packets in your kitchen and bathroom. Make a list of five things that are sold by volume (ml or fl.oz – fluid ounces) and five that are sold by mass (g or lb). Is it true that most runny things are sold by volume and the others by mass? Suggest reasons why this should be so. What about toothpaste, jam, and sauce?

8. 3.14159...

This is the strange number called *pi* (say 'pie'). Pi is the name of the Greek letter π, and the number keeps turning up in circles.

We don't usually use all those decimal places. 3.14159 . . . is very close to $3\frac{1}{7}$, or $\frac{22}{7}$. Almost all the time we can use $\frac{22}{7}$ for π.

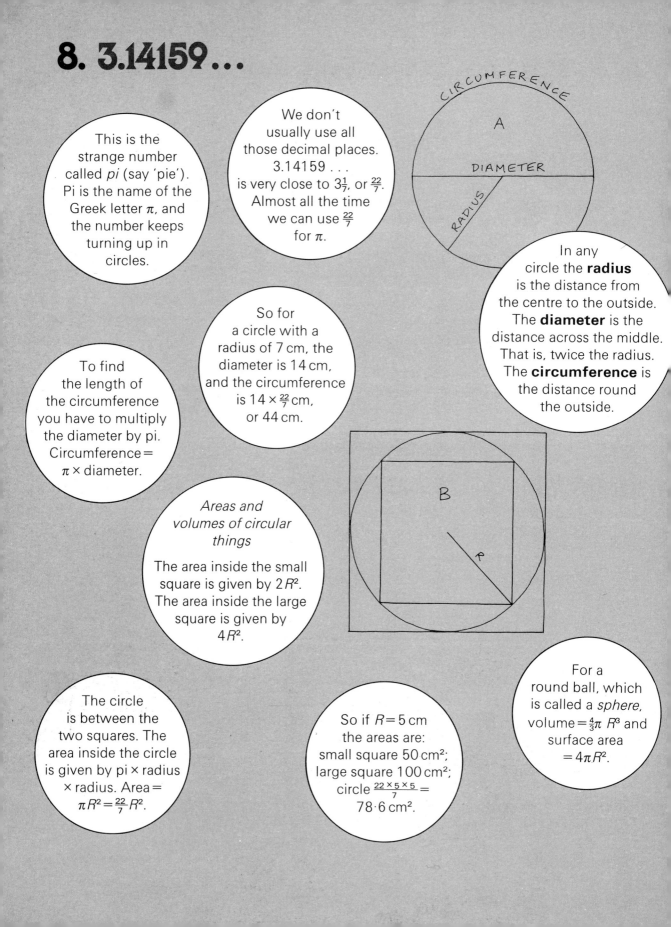

In any circle the **radius** is the distance from the centre to the outside. The **diameter** is the distance across the middle. That is, twice the radius. The **circumference** is the distance round the outside.

So for a circle with a radius of 7 cm, the diameter is 14 cm, and the circumference is $14 \times \frac{22}{7}$ cm, or 44 cm.

To find the length of the circumference you have to multiply the diameter by pi. Circumference = $\pi \times$ diameter.

Areas and volumes of circular things

The area inside the small square is given by $2R^2$. The area inside the large square is given by $4R^2$.

The circle is between the two squares. The area inside the circle is given by pi × radius × radius. Area = $\pi R^2 = \frac{22}{7} R^2$.

So if $R = 5$ cm the areas are: small square 50 cm²; large square 100 cm²; circle $\frac{22 \times 5 \times 5}{7} = 78 \cdot 6$ cm².

For a round ball, which is called a *sphere*, volume = $\frac{4}{3}\pi\, R^3$ and surface area = $4\pi R^2$.

Questions

1 A chapati has a radius of 7 cm. What is its area?
2 In question 5.2 you should have found that the circumference of your head was about 55 cm. This is about 22 inches. Suggest why hat sizes are about 7.
3 A farmer near Huddersfield keeps his cow Matterhorn in a square field which is 25 m along each side. One night all the fences are flattened by a stampede of hippos from Halifax. To stop the cow running away he ties her with a piece of string to a post in the middle of the field. The string is 14 m long. Does Matterhorn have more grass to chew or less than before the coming of the hippos? (*Hint*: work out the area of the square field, and then the area of a circle with $R = 14$ m.)
4 This roundword is about length, area, volume, and circles. Copy it into your book. Don't fill in the roundword in this book.

R = 7cm

Down

1 A thousand of these units of volume in a litre (2,2)
2 Water round the 4 of a castle (4)
6 Next whole number greater than π^2 (3)
7 Used for controlling a horse, but sounds like water from heaven (4)
9 Measure of the size of a surface, perhaps in cm² (4)
11 Number of dimensions in a flat surface (3)

Across

3 There is friction when two things do this (3)
5 Farmer's measure of 9; 10 000 m² (7)
8 Distance across the middle of a circle (8)
10 Chart useful for showing survey results (3)
12 Sounds like 22/7, but really another sort of chart (3)
13 Unit of force (6)

Clockwise

4 Distance round the outside of the circle (13)

9. How dense are you?

Which weighs more, a kilogram of feathers or a kilogram of gold? The answer is that they weigh exactly the same amount, for they both have the same mass, 1 kg. The difference is that a 1 kg bar of gold would be about the size of a Mars bar, but 1 kg of feathers would fill a huge pillow.

The gold packs its 1 kg mass into a much smaller volume than the feathers. We say that gold is much more **dense**. That doesn't mean gold is stupid! It means that a small volume of gold has a large mass.

Heavy in your hands

Pass round the class, if possible, some blocks of wood, iron (or metal weights), polystyrene or foam rubber, lead, and cork. Write down a list of these things in order of how heavy they feel for their size.

Polystyrene and cork feel light. Metals feel heavy. This is because metals are more dense.

The **density** of any stuff is the mass of one cubic centimetre. A volume of 1 cm³ of water has a mass of 1 g. The density of water is 1 gram per cubic centimetre, or 1 g/cm³.

The density of gold is 19 g/cm³, because 1 cm³ of gold has a mass of 19 g. Gold is much denser than lead, but not the densest stuff in the world. There is a poisonous metal called osmium which is even denser. But osmium doesn't look or sound as good as gold – no one has made a film called 'Osmiumfinger'!

Table of densities	
Empty space	0 g/cm³
Polystyrene	0.1 g/cm³
Cork	0.2 g/cm³
Wood	about 0.6 g/cm³
Water	1.0 g/cm³
Stone	about 3 g/cm³
Iron	7 g/cm³
Lead	11 g/cm³
Gold	19 g/cm³
Osmium	23 g/cm³

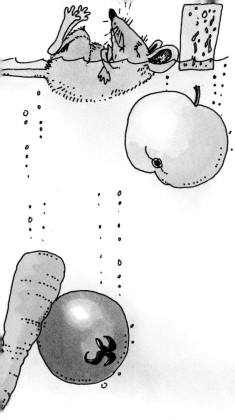

If sharks don't swim they sink

Anything will float in water if its density is less than the density of water, 1 g/cm³. Wood floats. Metal sinks.

Fruit and vegetables are mostly water, and they usually *just* sink or *just* float. Try this at home with a bowl of water. You will probably find that apples, oranges, and bananas float, but carrots and tomatoes sink. Oranges sometimes sink when they are peeled; the fruit is denser than the peel.

Animals are also mostly water. But animals have a layer of fat under the skin, and air spaces inside, like lungs. So most animals *just* float in water. When you lie back in a swimming pool your face should float above the surface. But fish have to stay under water. Their life would become difficult if they kept bobbing up like corks! Sharks are denser than water. If they don't keep swimming they sink to the bottom.

Questions

1 How dense are you? Make a guess, remembering that you just float in water.
2 Write down which of the following you would expect to float in water: gold, cork, iron, wood, orange peel. Explain why.
3 Mercury is a shiny, runny metal with a density of 14 g/cm³. Write down ten things you would expect to sink in water but float in mercury. What do you think it would feel like to jump into a swimming pool full of mercury?
4 Copy and complete this wordsquare, and write down exactly what the hidden word means.

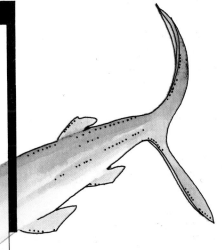

1 Metal with density of 19 g/cm³
2 Very dense things feel this
3 Dense things do this in water
4 This has to swim, or it sinks
5 The densest stuff of all
6 This has a density of 1.0 g/cm³
7 Ten times less dense than water

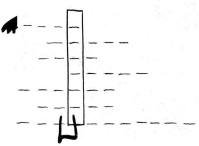

10. Eureka! or The incredible bulk

Once upon a time there was a Greek scientist who liked to think in the bath. His name was Archimedes (say 'Arky-meedeez'). He lived in Syracuse, and because he didn't have a bath of his own he used to go along to the public baths in town.

One morning he sat there worrying. The king had come to him the night before with a problem. 'Arky, you're a bit of a genius. You keep inventing pumps and things. How can I find out if my new crown is pure gold? I gave the royal crownmaker a block of pure gold, but I think he may have pinched part of the gold and mixed in some other metal to make up the mass. How can I find out? Better still, how can *you* find out?'

He wasn't as dense as he looked . . .

Archimedes scrubbed the back of his neck, deep in thought and soapy water. He could measure the mass of the crown. He could borrow from the king a block of pure gold of the same mass. Now if the crown were pure gold it would have the same volume as the block.

He could easily measure the volume of the block. But how could he possibly measure the volume of the fancy and elegant crown? And then he did something skullbendingly brilliant . . .

. . . He dropped the soap.

The bath was completely full. So when the soap slipped in, a slurp of water slopped over the side. Suddenly Archimedes knew how he could measure the volume of the crown. He

Eureka! Eugenius!

was so excited he jumped straight out of the bath, wearing nothing but a few bubbles. He ran all the way home, yelling at the top of his voice that he'd found the answer. 'I've found it' he screamed – only he screamed in Greek – *'Εὑρηκα'* (say 'Yoo-reeka').

What had struck Archimedes, apart from the cold, was that if you put a solid object into water it must push some water out of the way. Take a container and fill it to the brim with water. Put a solid object in. The water that spills over must have exactly the same volume as the object you put in.

You can use this idea to measure the volume of your thumb.
Fill a milk bottle to the top with water. Stick your thumb in as far as it will go.

ake your thumb out again.
ee how much water it takes,
om a measuring jug, to fill the
ottle to the top once more. That
the volume of your thumb.

. . . and nor was the crown

Archimedes used this method to measure the volume of the crown. The crown took up more space than it should. The volume of the crown was more than the volume of the pure gold block, even though they had the same mass. The crown wasn't pure gold. The king had been cheated.

Questions

1 Why was dropping the soap important?
2 Describe an experiment you could do to measure the volume of your own head.
3 Suppose you wanted to make a life-size model of your own head out of pure gold. How would you work out what mass of gold you would need? How much would it cost, if the volume of your head is 3000 cm³, and the price of gold is £10 per gram?

11. Oh buoy!

Why do big boats float? They are made mostly of iron and heavy metals. How do they float in water?

The short answer is **buoyancy**. Anything you put into water seems to weigh less than it does in air. Hang a 10p piece on a piece of cotton from a spring balance. Note what it weighs (about 0.125 N). Then hang it in water. You will find that the spring balance shows only about 0.111 N. The coin seems to have lost 0.014 N – 11 per cent of its weight.

What happens is that the water pushes upwards on the coin. As the coin sinks into the water, it pushes some water out of the way. This water is *displaced* by the coin. At the same time the weight of the coin seems to be reduced by the same amount as the weight of the displaced water. So the *upthrust* on the coin is the same as the weight of the water displaced.

Archimedes' Principle

Archimedes was an important scientist. He worked out the connection between the upthrust and the displaced water.

When an object is partly or totally immersed in water it feels an upthrust equal to the weight of the water displaced.

Archimedes's overweight daughter
Felt light when she got in the water,
  'Hey Dad, this is great.
  I'm sure I've lost weight,
Even though I ate more than I oughta!'

'My dear, from the truth you have drifted.
It's just that you *feel* uplifted;
  And the uplifting force
  Is equal, of course,
To the weight of the water you've shifted.'

'Why, even a heavyweight fighter
When he jumps in the bath will feel lighter.
  But the weight of the tide
  That flows over the side
Is all that's uplifting the blighter.'

We can often use Archimedes' Principle to work out the density of an object. The water displaced by the 10p piece had a weight of 0.014 N. So its mass must have been 1.4 g. So the volume of water displaced must have been 1.4 cm³, since the density of water is 1 g/cm³.

But the coin must have displaced exactly its own volume of water. (Remember Archimedes and the crown?) So the coin must have the same volume, 1·4 cm³.

So the density of the 10p piece must be $\dfrac{12\cdot5\,\text{g}}{1\cdot4\,\text{cm}^3} = 8\cdot9\ \text{g/cm}^3$ approximately.

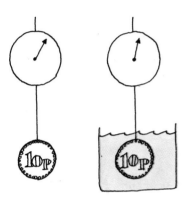

Why do ships float?

If a ship were made of solid iron it would sink. But in real ships the metal is spread out and wrapped round lots of air. There are cabins and cargo holds and engine rooms. Most of the volume of the ship is full of air. The result is that the density of the whole ship – metal and air together – is less than the density of water. So the ship floats.

The size of a ship is called its displacement, and is measured in tonnes. That is the mass of water displaced by the ship. It is also the mass of the ship. The ship is designed in a shape that displaces enough water to let it float, without the water pouring in over the top.

Questions

1 The *Titanic* had a displacement of 60 000 tonnes. What mass of water did she displace? What was the volume of the ship below the waterline when she set off on her maiden voyage on 10 April 1912?

2 A piece of cork has a mass of 10 g. What is its volume? (See **9** for its density.)
Cork floats in water. If you drop this piece in, it will feel an upthrust equal to its weight. What mass of water is displaced? What volume of water is displaced? What fraction of the cork is under water when it floats?

3 Your friend is about to go away for a holiday at the seaside. Salt water is denser than fresh water. Write a letter to your friend, explaining why it should be easier to float in the sea than in a swimming pool.

12. Perfect timing

The main railway line reached North Wales 150 years ago. For the first time passengers could travel all the way from London to Holyhead. Soon they noticed something odd. The journey east seemed to take 33 minutes longer than the journey west. *Always. Every time.* They could not understand it.

Why should this happen? Can you guess why the trains should run more slowly going from Wales to England than they did from England to Wales? Write down your suggestions.

The right time

How do you know what time it is? Many of us have super-accurate digital watches. They are guaranteed not to lose a second in a week. But do they all agree on what time it is? No. Usually they are all set at different times.

How do you know what time to get up in the morning? Does someone call you? Do you listen to the radio? Watch breakfast TV? What is the right time, anyway?

The idea that there should be a scientific standard of time began at Greenwich, just outside London. Scientists at the observatory there called this **Greenwich Mean Time**, or **GMT** for short. This became a standard for the whole world.

Every hour, on the hour, the BBC broadcast the Greenwich Time Signal. You can hear it on Radio 4 at 7.00, 8.00, 9.00, and so on. It sounds like six beeps, called 'pips'. The last pip is a bit longer than the others, and this sounds exactly on the hour. That's the most accurate way to set your watch.

But 150 years ago GMT was not in general use. The railway stations in north Wales set their time by the Craig-y-Don gun. Their time was 16·5 minutes behind Greenwich. They had no radio, telephone, or TV to warn them. So trains going east seemed to arrive 16·5 minutes late, and trains going west seemed to arrive 16·5 minutes early. That is why the journey seemed to take 33 minutes longer going eastwards.

Split seconds

The basic unit of time is the second. We often use longer units, but scientists need shorter ones too:

1 day = 24 hours
1 hour (hr) = 60 minutes
1 minute (min) = 60 seconds
1 second (s) = 1000 milliseconds
1 millisecond (ms) = 1000 microseconds

A year is about $365\frac{1}{4}$ days long. To make life easier, we make most years exactly 365 days, and give every fourth year one day extra. The years with 366 days are called leap years. In leap years February has 29 days instead of the usual 28. Leap years include 1988, 1992, 1996, and 2000. But 2100 will not be a leap year; only every fourth century counts.

Questions

1 Which will be the first three leap years after 2009?

2 Before modern clocks and watches had been invented, people used all sorts of devices for measuring time. The Romans used sundials by day. At night they burned candles with the hours written on the side. Other people used hour glasses with sand, or clocks based on dripping water. Look in books in the library. Find out about at least four of these clocks. Write five lines about each, with a drawing, to explain how they worked.

3 Describe in detail how you would make one of these:
   a A simple sundial. When would you expect the shadow to be shortest?
   b A candle alarm clock. (*Hint*: the flame might burn through a piece of string at the time you want to be alarmed!)
   c A water clock, using a kitchen spring balance. (*Hint*: fill the pan with water, and use a piece of paper towel to help it dribble over the side. Watch the weight of water.)

4 Copy and complete this clockword.

1 d This way the trains seemed to take longer (4)
2, 13, 6 clockwise. The world's first time standard (9,4,4)
2d is short for this (3)
3 d 5 d There's an extra day in these (4,5)
4 d 11 a Not new, and you won't find one on a digital watch (6,4)
5 d See 3 d

6 See 2.
7 a This golf ball support sounds like the beginning of time (3)
8 a Look! This tells you the time (5)
9 a Article found in 11 and 13 (2)
10 a The sands of time trickle through this glass (4)
11 a See 4.
12 d Short for ante meridiem (morning) (2)

13. How fast can you go?

The **speed** of anything is the distance it moves in one unit of time. A car goes 30 miles in one hour. Its speed is 30 miles an hour, or 30 miles per hour, or 30 m.p.h.

A motor-bike goes 12 miles through London in 20 minutes. Is it breaking the speed limit?

20 minutes is one third of an hour, since 20 × 3 = 60. So the motor-bike is doing an average of 12 miles × 3 = 36 miles in 1 hour. Its average speed is 36 m.p.h. If the limit is 30 m.p.h. all the way, then the motor-bike is breaking the limit.

Scientists usually measure speed in metres per second (m/s) rather than m.p.h. To work out the speed, divide the distance by the time taken:

R.I.P   R.I.P
A learner was driving his car
When his friend said,
'You're too slow by far.
If you drive at this rate
We are bound to be late
Drive faster!' He did, and they are!

$$\text{Speed} = \frac{\text{Distance}}{\text{Time}}$$

If your friend is on a bike and covers 100 metres in 10 seconds, then

$$\text{Speed} = \frac{100\,\text{m}}{10\,\text{s}} = 10\,\text{m/s}$$

She travelled at a speed of 10 m/s.

The speed of champions
You can probably walk fast at 4 m.p.h. (1·8 m/s), and run at 12 m.p.h. (5 m/s). This is what world record holders can do:

	Distance	Time	Speed	
Sky-diver	—	—	185 m.p.h.	83 m/s
Sprint cyclist	200 m	10·7 s	42 m.p.h.	19 m/s
Horse and jockey	1·5 miles	2.5 min	35 m.p.h.	16 m/s
Sprinter	100 m	9·9 s	22 m.p.h.	10 m/s
Swimmer	100 m	49·4 s	4·5 m.p.h.	2 m/s

The human body is not designed for swimming. You can run five times as fast as you can swim. But you can cycle twice as fast as you can run. That is because cycling is **efficient**. Very little work is wasted.

Your biggest muscles are above your knees, in your thighs. Put your hands flat on your thighs, put your feet flat on the floor, and push yourself to the back of your chair. You can feel your thigh muscles going tight under your hands. It doesn't matter whether you are a hefty weight-lifter or a feeble weakling; the muscles in your thighs will still be much bigger than any in your arms or the rest of your body.

When you run, most of the work of your thigh muscles goes into lifting you off the ground at each step. Not much is left to push you forward. Running on the spot is no easier than running 100 metres. But on a bike you can sit down. The bike takes your weight, and most of the work of your thigh muscles goes into pushing you forward. Little effort is wasted. High efficiency.

Velocity

Speed is the distance covered in a unit of time. **Velocity** is **speed in a particular direction**. Ride your bike north at 15 m.p.h. and your velocity north is 15 m.p.h. Your velocity south is minus 15 m.p.h.

A runner who does a 400 m circuit of a track in 80 s has a speed of 400 m/80 s = 5 m/s. But her average velocity is zero, because by the end of the circuit she's back at the start. The total distance she has moved in any direction is zero.

Questions

1 In each case write one sentence to explain the meaning of these words: *speed, velocity, efficient, speed-limit*.

2 You can tell how far away a thunderstorm is, using the speed of sound. When lightning strikes within sight of you, you see the flash in much less than one millisecond. The sound, which we call thunder, takes much longer to reach you.

   Sound travels at about 760 m.p.h., or *** m/s. This means that sound takes about * seconds to go one mile, or 3 seconds to go one kilometre. Fill in the ** above, and explain how you can find out how far away the storm is by counting the seconds between the lightning and the thunder.

3 Copy and complete this wordsquare. What is the mystery amount in town?

1 How far you have travelled
2 Speed at the end of triumph
3 5 in a particular direction
4 The number of these in 1 s is 3 or 5
5 **1/10** in m/s
6 This is much quicker than running . . .
7 . . . because it is more this
8 At this, 2 m/s is a very high 5
9 Here are your biggest muscles
10 Emit backwards

14. A change of pace

Have you ever been in an underground train? When it starts there is often a terrific jolt. People standing up have to hang on, or they are likely to fall over.

You can feel the same effect in a car but it is less violent. When the car sets off from the lights you feel yourself being pressed back into the seat. You feel this because of **acceleration**.

Acceleration = rate of change of velocity

The lights go green. The driver puts his foot down. The car zooms off. The velocity of the car goes from 0 to 30 m.p.h. The car is accelerating. That is why the foot pedal on the right is called the accelerator. Press it, and the velocity of the car will change.

A tennis ball comes at you at 5 m/s. You swing your racket and bash it back at the same speed. The speed of the ball is the same, but its velocity has changed, from 5 m/s to −5 m/s. The racket has accelerated the tennis ball, and changed its velocity by 10 m/s.

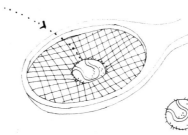
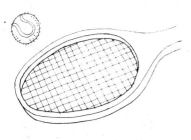

The units of acceleration are velocity per unit of time. Suppose the racket touched the ball for 0·1 s. The change of velocity was 10 m/s in 0·1 s, or 100 m/s per second. We say 100 metres per second per second, or 100 m/s².

Slowing down

The lights go red. The driver puts on the brakes. The car screeches to a stop. You feel yourself thrown forward in your seat. Again you are feeling acceleration. But this time the velocity is getting less. The acceleration is negative. This is often called deceleration.

Newton and the force

Isaac Newton was a scientist. He lived 300 years ago in the fens of Lincolnshire and Cambridge. One of the things he worked out was the link between force and acceleration.

Force = Mass × Acceleration $F = ma$

When the car accelerates you feel pushed back. But you are not pushed back. The car is accelerating *forwards*, away from you. You will be left behind unless you are pushed forwards.

The car seat pushes you forwards. This is the force. The push from the seat gives you the acceleration you need to keep up with the car.

One thing that makes Superman special is that he can accelerate to immense speed, without using force. Just by lifting an arm he can zoom into the sky. The rest of us aren't so lucky. We have to use force every time we want to change our velocity.

Questions

1 Copy and complete the following sentences:
   a Acceleration is the . . . . of . . . . . . of velocity.
   b Acceleration can be measured in metres per . . . . . . per . . . . . .
   c When the car accelerates you are . . . . by the car seat.

2 A sports car sets off due north. Its velocity goes from 0 to 30 m/s in 10 seconds. Then the driver stamps on the brakes, and stops the car in 6 seconds. What is the acceleration in each part of this journey?

3 At the start of the hundred metres, the sprinter leans forward. Feet firmly on the ground, or the blocks. The gun goes off. Both legs straighten rapidly. This applies forwards force to the body. Suggest why this is useful in reaching a high velocity.

4 Your cousin is going to travel in an underground train for the first time. Write a postcard to explain why it is important to hold on when it starts. Don't forget the force is always with you.

5 Write down whether you think Newton would have believed in Superman. Explain your answer.

15. Staying in orbit

A space lab is zooming round the Earth. Every 90 minutes it does a complete orbit, more than 40 000 km. The motors have been switched off. The lab is 300 km above the ground. Why doesn't it fall down? Write down what you think keeps it up.

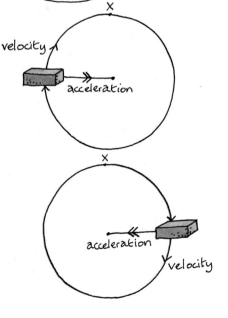

Tie a roll of sellotape on a chain of three strong rubber bands. Whirl it round your head. Make sure you don't let go. What happens to the rubber bands when you whirl fast?

Circling

Here is a brick flying round in a circle. At this moment it is moving north, up the page. But when it gets to X it will be moving to the right – east. Later it will be going south, and west. The direction is changing all the time.

The speed of the brick stays the same. But its velocity is always changing. *Remember*, velocity is speed in a particular direction. Because the direction is changing, so is the velocity of the brick. Change in velocity means acceleration.

This brick is accelerating towards the centre of the circle. It doesn't get any closer to the centre, but it has to accelerate all the time to stay the same distance away. Just like the Red Queen in *Alice in Wonderland*. She had to run faster and faster to stay in the same place.

Anything that moves in a circle is accelerating towards the centre.

What causes acceleration? Force. Why does anything move in a circle? Because a force is pulling it in.

Danger – high tension

What would happen if you cut the rubber bands? The sellotape would fly across the room. What keeps the sellotape going round is the pull from the rubber band. The force of tension in the rubber band makes the sellotape accelerate towards the centre.

snip!

Some people say that the sellotape pulls outwards with *centrifugal force*. That isn't really what is happening. *You* have to pull the sellotape in. *You* have to make tension in the rubber band. *You* have to provide the force that keeps the sellotape accelerating inwards. *Your finger* is what produces the force – not the sellotape pulling outwards.

In the space lab, the astronauts and the coffee seem to be *weightless*. They are accelerating towards the Earth all the time. So is the whole lab. The force of gravity is just enough to keep them in orbit. They feel exactly as if they are falling, all the time. This is *weightlessness*.

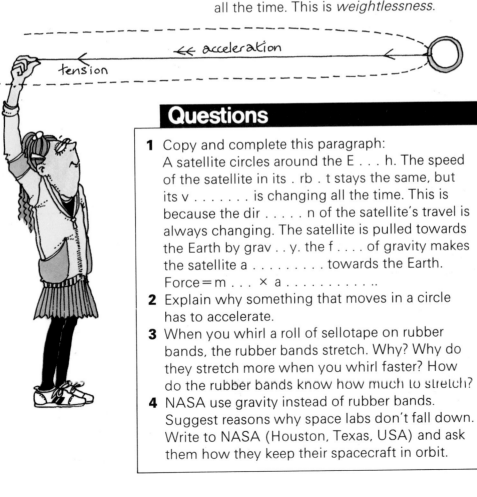

Questions

1 Copy and complete this paragraph:
A satellite circles around the E . . . h. The speed of the satellite in its . rb . t stays the same, but its v . . . . is changing all the time. This is because the dir . . . . . n of the satellite's travel is always changing. The satellite is pulled towards the Earth by grav . . y. the f . . . . of gravity makes the satellite a . . . . towards the Earth.
Force = m . . . × a . . . .

2 Explain why something that moves in a circle has to accelerate.

3 When you whirl a roll of sellotape on rubber bands, the rubber bands stretch. Why? Why do they stretch more when you whirl faster? How do the rubber bands know how much to stretch?

4 NASA use gravity instead of rubber bands. Suggest reasons why space labs don't fall down. Write to NASA (Houston, Texas, USA) and ask them how they keep their spacecraft in orbit.

16. Bouncing back

A ping-pong ball drops from 50 cm on to a piece of wood. It bounces down on to the table, and it bounces away. The whole action takes 1.5 seconds. What can we find out by looking at it in detail?

First look at the drop. The ball falls straight down, towards the centre of the Earth. Gravity always pulls straight down. In fact that is what we mean by *down* – the way that gravity pulls.

PARABOLA

The separate images show the ball every 0·03 s – that is every 30 milliseconds. Can you see that the ball drops more in the second 30 ms than in the first – and more in the third 30 ms than in the second? Remember gravity is pulling the ball down. Its weight is a force, and when a force acts on a mass it causes acceleration. The ball is falling with the acceleration due to gravity. As long as the ball falls it drops faster and faster – until it hits the wood at the bottom.

Clunk!

When the ball touches the wood it doesn't stop dead. First it has to slow down until its down velocity is zero. Then it has to start moving upwards. For all the time that it is touching the wood the ball is accelerating upwards.

As it leaves the wood it is moving up at almost the same speed as it was falling down before. But not quite. You can see that the images of the ball are a little bit closer together as it goes up than they were when it was coming down.

You can see also that the ball does not bounce as high as the place it fell from. No ball ever bounces all the way back. If it did it would go on for ever. The second bounce is always lower than the first.

Which way does it go?

The piece of wood is tilted at an angle to the table. So the ball bounces out to the right. The shape of each complete bounce is called a **parabola**.

When the ball hits the table it bounces off at nearly the same angle. Here it hits, and takes off, at about 70°.

Questions

1 How long does the ball take to complete the first bounce, from wood to table? (*Hint:* how many images can you count?) Would you expect the second bounce to be shorter? Is it?
2 Measure with your ruler on the page the height from which the ball was dropped. Also the height to which it bounced. Divide the bounce height by the drop height to get a measure of how bouncy the ball is on wood. It should come to between 0·5 and 1. Do you get the same result for the second bounce?
3 As the ball rises from the table, the images get closer together. Can you explain why?
4 A table tennis net is 15 cm high. What is the minimum height you would have to drop a ball from if it is to bounce over the net (a) on the first bounce, (b) on the second bounce?
5 Describe a fair experiment that you could do to measure bounciness. You could try a variety of conditions, including tennis balls on grass, and meatballs on spaghetti.

17. Under pressure

When you push your thumb against a notice board, your thumb does not sink in. But if you use your thumb to push a drawing pin, the drawing pin sinks into the board. Why should this happen?

When you push against the board with a certain force, the board pushes back with the same force. Your thumb doesn't sink in. But the sharp end of the drawing pin does.

That's the point . . .

When you use the same force to push the drawing pin against the board, the board is not strong enough to resist. The point sinks in because the force is concentrated. All the force is focused in that narrow point. That means a high **pressure** at that point.

Pressure is force per unit area

Suppose you push with a force of 10 newtons.

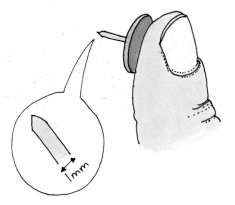

Area of end of thumb
$= 2\,\text{cm} \times 2\,\text{cm}$
$= 4\,\text{cm}^2$

Area of point of drawing pin
$= \pi R^2$ (area of a circle)
$= \frac{22}{7} \times 0\cdot05\,\text{cm} \times 0\cdot05\,\text{cm}$
$= 0\cdot008\,\text{cm}^2$

Pressure = force/area

$$\text{Pressure} = \frac{10\,\text{newtons}}{4\,\text{cm}^2}$$
$$= 2\cdot5\,\text{newtons/cm}^2$$

$$\text{Pressure} = \frac{10\,\text{newtons}}{0\cdot008\,\text{cm}^2}$$
$$= 1250\,\text{newtons/cm}^2$$

So if you use a drawing pin you are putting 500 times more pressure on the board than if you just use your thumb. No wonder the pin goes into the board. It can't take that sort of pressure! *Remember*, you are pushing the drawing pin with the same force. But the area of the point is small. So the pressure is huge.

The ship of the desert

For thousands of years desert tribes have used camels to help them travel across the sands. The main reason for this is that camels can survive for many days without food or water. But camels also have big feet.

Animals with sharp hooves sink into the sand. This makes walking hard work. Camels don't sink in. Why not? Their feet have a large area; so the pressure on the sand is low. Just as your thumb doesn't sink into the notice board, so the camel's feet don't sink into the sand. The camel seems almost to float on the sand, like the ship of the desert.

Questions

1 Suggest a reason why some people in cold countries use big, wide snowshoes. How does pressure come into it?

2 An elephant weighs 40 000 newtons. Each of its feet has an area of 500 cm². If it stands on one foot, what pressure does it put on the ground?

   A lady weighs 500 newtons. Her stiletto heel has an area of 0·5 cm². If she puts all her weight on one heel, what pressure does she put on the floor?

   Why do you think stiletto heels are banned in many public places? Would you rather have your toe stepped on by an elephant, or by a lady in stiletto heels?

3 Copy and complete the wordsquare. Write down the meaning of the hidden words in the box.

Elephants welcome but NO stilettos please.

1 Rate of change of velocity
2 The force of this pulls you down
3 Things in orbit accelerate towards it
4 The camel is the . . . . of the desert
5 Banana skins don't give you much
6 Speed = distance/this
7 Speed in a particular direction
8 Mass per unit of volume
9 Pressure = force per unit of this
10 The amount of stuff in an object
11 Force per unit area
12 Where GMT began
13 Distance across a circle
14 Force that pulls a string tight
15 Mass × acceleration
16 Its area is πR^2
17 The shape of a bounce
18 There's an extra day in this
19 'I've got it!' in Greek

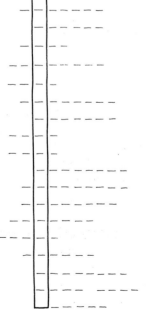

18. Sizzling sausages

Sizzling sausages, boiling water, burning fire. These things are hot. Snow and ice are cold. We all know that, but scientists need to know how hot or how cold. And so they measure the **temperatures** of these things.

Celsius temperatures

Ice melts at zero degrees Celsius (0°C).

Water boils at one hundred degrees Celsius (100°C).

These are called the **fixed points** of the Celsius temperature scale. Anything hotter than boiling water has a temperature above 100°C. Anything colder than melting ice has a temperature below 0°C.

A healthy human body has a temperature near 37°C. This is not a fixed point, but most people have temperatures close to 37°C. If you have a fever your temperature may go up to 39°C, but if it rises above 41°C you are practically dead.

If you get very cold your temperature may drop a few degrees. But you are in danger if your temperature gets as low as 30°C.

Thermometers

Thermometers measure temperature. With a thermometer you can find out how hot something is. A bulb thermometer, like the one shown here, is made of glass. The bulb at the bottom is filled with either silvery mercury or red-coloured alcohol.

To measure the temperature you should put the thermometer in whatever you are measuring. Part of the liquid in the bulb rises up the tube – the *stem* of the thermometer.

As the liquid inside gets hotter it expands. In other words it takes up more room. The result is that it climbs higher up the tube. The higher it climbs, the higher is the temperature it indicates.

You read the temperature at the top of the liquid from the scale printed on or by the stem. The thermometer here is showing a temperature of about 81°C.

But temperature isn't everything

When you eat hot jam pudding, you may burn your tongue on the jam. Never on the pudding. The whole pudding, jam and all, has been together in the oven. So it must all be at the same temperature. So why do you burn your tongue only on the jam?

The answer is that temperature isn't the only thing that matters. You also need to know that jam can hold more **heat** than the rest of the pudding. We'll find out more about heat in the next chapter.

Questions

1 What would you use to measure how hot something is?
2 Write down and complete this sentence:
The fixed points on the Celsius temperature scale are . . .
3 From the picture, what would you expect to be the temperature of a packet of peas which you took out of the freezer?
4 The temperature of the air is measured in the shade – where it is not warmed by the sun. Find out the highest and the lowest temperatures ever recorded on Earth.

19. Getting warmer

"Lor Lynne, it's perishin' cold today. I can feel all that temperature running out through my feet into the snow."

"George, temperature doesn't move. Temperature tells you how hot things are. The snow is minus a few degrees, and your feet should be about 30 degrees."

"That's what I'm beefing about. That cold comes freezing up through my boots...."

"No no George. Cold doesn't move. And degrees don't move either. It's only heat that moves. Wrap your hands round this cuppa and feel the heat running in. When the heat runs in that's what puts up the temperature."

Temperature and heat

On a cold day in winter you hold out your hands to warm them in front of the fire. What comes from the fire is **heat**. The heat warms your hands. Their temperature goes up.

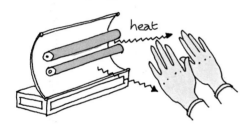

heat

If you go out to make snowballs your hands get cold. This is because the heat runs out from your hands into the snow. Some snow may melt. The temperature of your hands drops. They feel cold.

Heat always goes from hot things to cooler things. The heat goes from the hot fire to your cold hands. The heat goes from your warm hands to the cold snow.

> *Remember*, temperature is a measure of how hot something is. To change the temperature you need to move some heat. Nothing can get hotter unless more heat comes in. Nothing can get colder unless heat moves out.

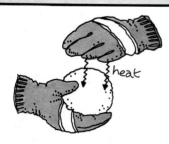

heat

We measure temperature in degrees Celsius. We measure heat in joules or calories. To heat 1 g of water by 1°C takes 1 calorie or 4·2 joules. To boil enough water to make a cup of tea takes about 50 000 joules or 50 kilojoules (kJ).

Travelling heat

We can travel by bus, car, train, or in many other ways. Heat has only three ways to travel:

1 Conduction. Heat flows directly from a hot object to a cold one. Wrap your cold hands round a warm cuppa, and the heat is conducted from the tea into the cup and from the cup into your hands. This is **conduction**.

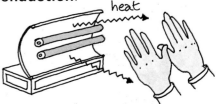

2 Radiation. Hold your hands in front of the fire and you can feel the heat. You don't have to touch the fire! When an object is very hot, heat will flow out from it in the form of radiation. This is called **radiant** heat.

3 Convection. To warm up a room, such as a classroom, you have to warm the air in the room. A hot radiator warms the air around it, and this will rise up. You can always feel the air warm above a hot radiator. As the air moves round the room the whole room is gradually warmed. This is called **convection**.

Questions

1 Copy and complete this sentence:
Heat can travel in three ways: 1 . . . 2 . . . 3 . . .
2 Snowballs feel cold. The sun feels warm. Explain how snowballs and the sun can change your temperature. How does the heat travel in each case?
3 Copy and complete this hot wordsquare.
Write one sentence to show what you understand about the word in the box.

1 Anything with a higher temperature is this
2 You need more of this to make anything 1
3 These degrees are used to measure temperature
4 The sort of heat we get from the sun
5 Snow and ice do this at zero 8,3
6 Heat will flow from hands into balls of this
7 Silver liquid, dense, and found in thermometers
8 These are used in measuring temperature
9 A measure of how hot a thing is
10 Your 9 goes up if you have one of these
11 With your hands you can feel 4,2 from this

20. The three bears

Once upon a time there were three bears. They lived in a house in the forest. One morning they went out for a walk while their porridge cooled, and a pesky child called Goldilocks crept into the house.

With 'bear-faced' cheek she tried their chairs. Father Bear's chair was too big and too hard for her. Mother Bear's chair was too soft for her. But Baby Bear's chair felt just right and very comfortable. Goldilocks sat on it so hard that the bottom fell out.

Then she tried the porridge. She found Father Bear's porridge too hot for her – and too salty. Perhaps he was Scottish, since true Scots always eat their porridge salty. She found Mother Bear's porridge sweet, but too cold. Baby Bear's porridge tasted great, and Goldilocks scoffed the lot.

Not content with messing about downstairs, Goldilocks went upstairs for a nap. One bed was much too hard; one was much too soft. Surprise, surprise, one bed was just right, and off she went to sleep. She didn't even hear the bears come back.

They weren't exactly pleased when they found the porridge.

And when they went upstairs they were a little upset to find that their beds had been slept in. They woke Goldilocks and chased her out of the house, and she didn't stop running till she got home.

Questions

1 The bears all had different chairs. Do adult *people* usually have different chairs? What happens in your house?

2 Goldilocks fitted into Baby Bear's chair, and found it comfortable. But under her weight it broke apart. What does this tell you about Goldilocks's weight? Was it greater than Baby Bear's? Were they about the same size ? What can you say about Goldilocks's density?

3 Scientists from the Porridge Institute at John o'Groats have measured the rate at which porridge cools. Here are their results.

Rate of cooling (°C/min)	0·90	0·83	0·62	0·50
Temp. (°C) above the room	59	55	44	38

Draw a graph of the rate of cooling (up the y-axis) against the excess temperature (x-axis) – the difference in temperature between the porridge and the room.

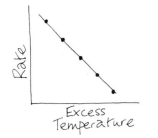

You should get something close to a straight line. This means that the rate of cooling is proportional to the excess temperature. This is Newton's Law of Cooling. What does this mean, in simple terms that even Goldilocks would understand?

4 A bowl of porridge cools much more slowly than a bowl of water. But most of porridge is water; so the thermal capacity must be about the same. That is, there must be about the same amount of heat in a bowl of porridge and a bowl of water. Suggest reasons why the heat should be lost more slowly from the porridge.

5 Would you expect porridge to cool more quickly in a mug, or in a wide, shallow bowl? Why?

6 Goldilocks was lucky to have a choice of beds to sleep in. There was an ancient Greek called Procrustes, who had a most unfriendly attitude towards visitors who didn't fit in his spare bed. Find out from the library what he did to them.

7 Mother Bear's porridge was sweet. She probably had sugar in it. Write her a letter, saying what dangers there might be in eating too much sugar.

21. Why does a kettle sing?

Have you ever listened to your kettle? Next time someone at home puts the kettle on to make a cup of tea, listen to the noises it makes. You get almost the same noises when you heat water in a beaker in the classroom.

Soon after the water begins to warm up you begin to hear a faint hissing. The hissing deepens to a dull roar. Look in while it is roaring. You will see that there are lots of bubbles at the bottom of the water, but not many at the top. This roaring is what people mean by the *singing* of the kettle. This is the noise that is hard to explain.

After a few minutes the roar dies down. As the water begins to boil the roar changes to the sloshy, bubbly noise of boiling.

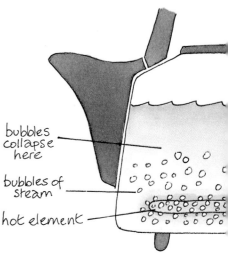

bubbles collapse here

bubbles of steam

hot element

LIQUID

Liquid, solid, and gas

Water is runny. It fills up the corners in the bottom of any container you pour it into. It runs out through any holes in a container. It feels wet on your fingers. It soaks up on to a handkerchief, a towel, or a piece of paper tissue. It doesn't sit around in lumps. Runny things like this are called **liquids**.

Put some water in the freezer and you can turn it into ice cubes. Ice cubes behave quite differently from water. Ice cubes are not runny. If you put some ice cubes into a container they don't fill up the corners. They don't run out through little holes. They don't soak up on handkerchiefs. They do sit around in lumps. They feel wet only if your warm hands melt the outside. Hard things like ice are called **solids**.

SOLID

GAS

Boil water and you turn it into steam. Steam is less dense than liquids or solids. It doesn't sit around in buckets or kettles. It can get everywhere. It can fill the whole room. Steam is a **gas**.

You can't feel gases. They don't have edges, and they aren't wet. You can't see most gases. But sometimes you can smell a gas. Every smell you smell is a little bit of gas getting up your nose. Fresh bread, spicy cooking, and sweaty socks all smell quite different. But all smells are gases. Gases fill bubbles and balloons.

The song of the kettle

The kettle sings with a dull roar. You can see lots of bubbles at the bottom, but none at the top. That is the clue to the noise.

As each little bit of water gets heated at the bottom, it turns into a bubble of steam. Those are the bubbles you can see. The bubble of steam is less dense than the water. So it floats up. But there it runs into cool water. The cool water takes heat away, and turns the steam back into water. The bubble collapses – almost as if it was pricked with a pin.

You can see hundreds of bubbles at the bottom. Each one collapses with a tiny BANG. All those tiny BANGs are what makes the dull roar. That is why the kettle sings.

Questions

1 Write down this list in your book, and beside each thing write down solid, liquid, or gas.
 a Air **b** Banana milk shake **c** The North Sea **d** North Sea Gas **e** Sand **f** String **g** The smell of onions **h** Custard.

2 What are bubbles full of? What is the outside made of?

3 Lemonade and other fizzy drinks have a gas in them called carbon dioxide. This gas starts bubbling out when you open the bottle or can. Suggest reasons why the bubbles rise towards the surface, and why they do not collapse on the way up.

4 Solid, liquid and gas are the **three states of matter.** At home you can find many things which are in two of these states. A tin of beans has solid beans and liquid sauce. A sponge has solid foam and gas bubbles. Choose five things, like these examples, and write a sentence about each to say which states they are.

5 Are *you* solid, liquid, gas or a mixture?

Isn't this a gas!

22. Early morning dew

Early in the morning the grass is often wet. Walk on it, and your feet get soggy. You leave trails of dark footprints. This wetness is called **dew**, and it comes out of the air.

On a warm day the heat from the sun raises the temperature of the ground, and the ground begins to dry out. Some of the water in the ground goes off into the air. Damp clothes hung on a washing line will dry in the sun, because the water goes off into the air. We say that the water **evaporates**.

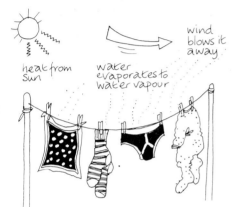

heat from sun

water evaporates to water vapour

wind blows it away

By the evening there is a lot of water in the air. This is steam – water gas – but we usually call it water **vapour**. Vapour is gas that is almost ready to turn into liquid.

How vapour vanishes

Water vapour stays in the air as long as the air stays warm. But at night the sun goes down. Everything begins to cool. As the air gets cooler it cannot hold so much water vapour. Tiny drops of water are formed from the vapour that was in the air. The vapour **condenses**, and turns into liquid water.

If the temperature goes on falling, more water vapour condenses. The drops of water may grow big enough to see. When they form on the grass we call them dew. Look carefully at the grass on a dewy morning. You will see that every blade is covered with tiny drops of water. Sometimes the grass looks almost grey.

A change of state

Solid, liquid, and gas (or vapour) are the **three states of matter**. They have various ways of getting from one state to another.

- Solids **melt** to turn into liquids. Snow melts to water in your hands.
- Liquids **freeze** to turn into solids. Water turns into ice in the freezer.
- Vapours and gases **condense** to make liquids. Dew forms on the grass.
- Liquids **boil** or **evaporate** to make vapours and gases.

Usually we say *boil* when we deliberately heat a liquid. We boil water in the kettle to make tea. But when water goes into the air from damp washing in the sun we say that the water *evaporates*.

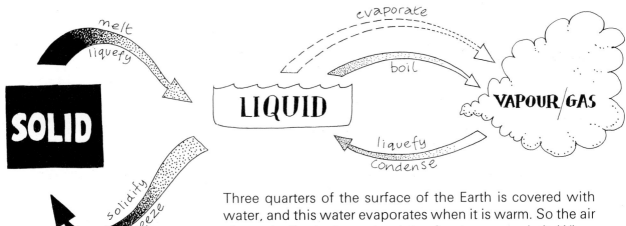

SOLID — melt / liquefy — LIQUID — evaporate / boil — VAPOUR/GAS

solidify / freeze

liquefy / condense

Three quarters of the surface of the Earth is covered with water, and this water evaporates when it is warm. So the air above the Earth always has lots of water vapour in it. When the air cools down low near the ground, the water condenses as dew. When the air cools high up in the sky the water may condense into rain drops. But more often it condenses into tiny little drops of water that are too small to fall down. They hang about in the sky, and we call them clouds.

Questions

1 Explain how the following turn from one state of matter to the other. **a** Ice into water. **b** Water into water vapour. **c** Water vapour into water. **d** Water into ice.
2 Suggest reasons why clothes on a washing line dry more quickly when the sun shines and when the wind blows.
3 Can you explain what fog is? Why does it usually disappear when the sun shines?
4 Copy and complete this crossword.

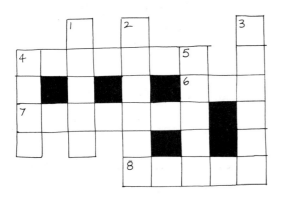

Across
4 When the kettle is singing, it really sounds more like this (7)
6 There is friction when two surfaces do this together (3)
7 Water isn't runny when you've frozen it into this (3,4)
8 Steam bubbles rise in the kettle because they are less ***** than water (5)

Down
1 Bubbles and balloons are full of them (5)
2 This is what you will get when you melt a solid (6)
3 A little bundle of gas, wrapped in liquid (6)
4 This may fall from the sky if lots of water vapour condenses (4)
5 Dew may turn grass grey instead of this normal colour (5)

23. The air that we breathe

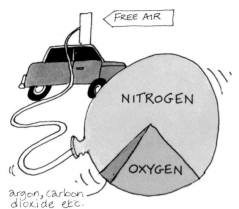

FREE AIR

NITROGEN

OXYGEN

argon, carbon dioxide etc.

Sparrows fly about in it, balloons and tyres are full of it, and we can't live without it. But what is air? Air is a gas – or to be more accurate, air is a mixture of gases.

You can see from the pie chart that air is almost completely made up from nitrogen (nearly 80 per cent) and oxygen (20 per cent). You can't see either nitrogen or oxygen, nor can you smell them or feel them. And that is why we hardly notice the air. But remember, if it wasn't there we wouldn't be able to breathe.

You want to stay alive?

What really matters to us is oxygen. Oxygen is the gas we need to stay alive. Without oxygen we could not burn coal and oil in fires and power stations. Fuel burns in oxygen to give energy. When we get oxygen into our bodies we can use it to get energy from the food we eat.

Nitrogen doesn't do much. It is **inert**. That means that it can't help to provide energy. Stuff won't burn in nitrogen; only in oxygen. You breathe nitrogen in, and then you breathe it out again unused. But oxygen is different.

When you breathe oxygen into your lungs, some of it goes on into the blood. That is how you get oxygen into your body.

Carbon dioxide is a waste gas that you breathe out at every breath. In the gases you breathe out there may be as much as 4 per cent of carbon dioxide. But you hardly breathe in any carbon dioxide at all, since in the air altogether there is only about 0.03 per cent.

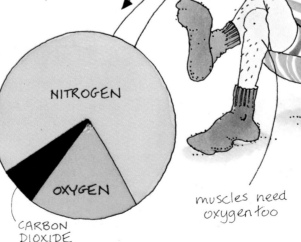

brain needs oxygen

nose picks up minute quantities of smells

air breathed in

NITROGEN

OXYGEN

CARBON DIOXIDE

muscles need oxygen too

lungs get air into blood

Air in funny places

You can't survive if you don't get enough oxygen. That is why some people have to carry tanks of air or oxygen with them. Divers who want to stay under water need to breathe. They carry bottles of air or oxygen.

Mountain climbers need oxygen at high altitudes. High above sea level the air is much less dense. On very high peaks the air is too thin to give the climbers enough oxygen to breathe comfortably.

The other stuff in the air

Air has some solid things in it – planes and sparrows and bits of dust. It also has some liquids, such as drops of rain. But almost all of it is gas. As well as nitrogen and oxygen there is argon, a tame, quiet gas even more inert than nitrogen. There is a little carbon dioxide. There is some water vapour. And there are all the things we call smells.

Although some things smell horribly strong, the amount of gas causing the smell is usually tiny – often as little as a millionth of one per cent. So you can smell a bad egg long before you can begin to measure how much bad egg gas there is in the air.

All together, the oxygen and nitrogen and argon and all the other stuff are called the **atmosphere**. That is the name for the air round the Earth.

Questions

1 Write down the names of the two commonest gases in the atmosphere. How much of each is there?
2 Apart from oxygen and nitrogen, make a list of three gases found in the air.
3 Helium is used in balloons because it is an inert gas. What do you understand by the word inert?
4 Find out from books in the library how we get energy from food and oxygen.

24. How can a candle burn?

Candles are made of wax. Solid wax won't burn. Cut a piece of wax from a candle and try to light it. You have no chance. The wax will melt, but it will not catch fire. So how can a candle burn?

Watch out for the wick

Up the middle of a candle is a piece of string. This is called the wick. Without the wick the candle won't work. Does that mean that when the candle burns it is really just the string on fire? No; string on its own burns quite differently. So the string and the wax somehow work together to make the candle flame.

Look closely at the burning candle, and see what is going on. Soon after the candle is alight, the top of the solid wax forms a cup. The wick comes up from the middle of the cup. The flame starts near the top of the wick. The flame itself is bright and white near the top, but bluish and dark below. Notice particularly that the flame is perhaps a centimetre away from the solid wax.

Do you see also that in the wax cup, round the bottom of the wick, is a little pool of liquid wax? Now we can work out what is happening.

vapour burns

here liquid wax begins to evaporate

heat from flame melts wax

liquid wax soaks up wick

cool air rushing up to flame keeps outside solid

It's only vapour that burns

The heat from the candle flame radiates down and warms the wax below. This melts the wax at the top of the candle. But liquid wax won't burn, any better than solid wax. That is where the wick comes in. We know that absorbent solids will soak up liquids. A handkerchief or a piece of paper towel will soak up spilled ink or milk. In just the same way string soaks up liquid wax.

Once the wax has been melted by the heat from the flame, the string can soak it up. So liquid wax flows slowly up the wick.

But how does that help? Liquid wax still doesn't burn. Before wax can burn it must turn into vapour, and mix with oxygen in the air. Then it can burn.

At the top of the wick is the flame. The liquid wax climbs up the wick and gets hotter and hotter, as it gets closer and closer to the flame. The liquid evaporates, and wax vapour streams off the wick into the air. It mixes with oxygen from the air, and it burns. That is the candle flame. The wax vapour burning in the air.

Seeing the vapour

Blow the candle out with a sharp puff, and you can see the vapour. The hot wax is still flowing up the wick, and for a few seconds you can see the white vapour flowing off the top of the wick into the air. It looks like smoke, but it smells quite different. And you can show it is wax vapour.

Blow the candle out. Wait about five seconds for the air to settle down. Then you will have a good stream of vapour flowing up. You can light this at the top with a match. The flame will run down the stream of vapour and back to the candle wick. A flame that leaps through the air.

Questions

1 What sort of wax burns? Solid, liquid, or vapour?
2 Write down why the wax melts at the bottom of the wick and why it vaporizes at the top.
3 Suggest reasons why the solid wax forms a cup at the top. It seems to melt in the middle more easily than at the outside. Can you explain why?
4 Find out how candles are made.
5 You can make a candle from a few cm of string, a milk-bottle top, and a dollop of one of these: cooking oil, baby oil, butter, or margarine. Write down what you try, and how they work. You will find that some things are better than others. Butter tends to crackle, for example.
CAUTION:
✱ DO NOT DO THIS EXPERIMENT AT HOME.
✱ BE VERY CAREFUL WITH FIRE.
✱ DO ALL EXPERIMENTS ON A SURFACE THAT WILL NOT BURN.
✱ MAKE SURE A TEACHER IS SUPERVISING.

25. No lumps

Sea water and swimming pool water taste terrible. Water from the tap doesn't taste of anything much, but the water in pools and the water in the sea both taste strongly, because of the stuff in the water.

Sea water is salty. You can taste the salt in the water. If you swim in the sea and then lie in the sun to dry, you can sometimes see the white salt powder on your skin.

Swimming pools have chlorine in the water. Chlorine is a disinfectant added to kill the germs. People who swim don't want to get ill from germs in the water. The chlorine is put in to protect people by killing any germs that might get in.

There is only a tiny amount of chlorine in the water – less than two parts of chlorine in a million parts of water. But chlorine is strong stuff, and even this is enough to taste, and to hurt your eyes if you get a lot of water in them.

You can't see the chlorine in the pool. You can't see the salt in the sea. Both the chlorine and the salt mix completely in to the water to make **solutions**.

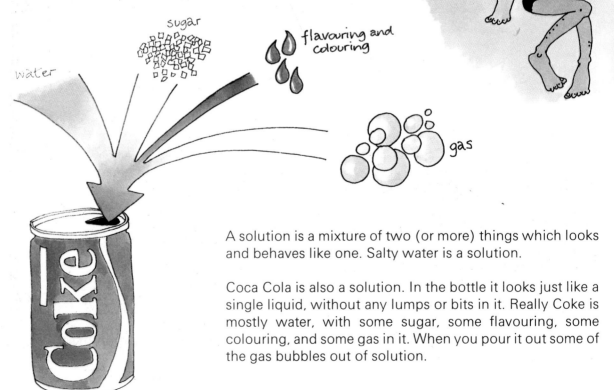

A solution is a mixture of two (or more) things which looks and behaves like one. Salty water is a solution.

Coca Cola is also a solution. In the bottle it looks just like a single liquid, without any lumps or bits in it. Really Coke is mostly water, with some sugar, some flavouring, some colouring, and some gas in it. When you pour it out some of the gas bubbles out of solution.

How to make a solution

You make a solution by **dissolving** one thing in another. For example you can stir a teaspoonful of sugar into a glass of water. As you stir the sugar seems to get less and less. Gradually it disappears, until there are no lumps.

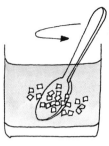

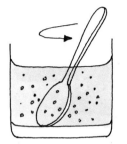

Look through a magnifying glass and you will see the little bits of sugar disappearing. They go soft at the edges and get smaller and smaller. When the lumps are gone, you have a solution.

After you have stirred vigorously for a minute or two there is no solid sugar left. You have made a solution of sugar in water. It looks just like pure water. There are no lumps. And it's difficult to get the solid sugar back again without drying the water off.

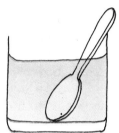

Dissolving and melting are different

	Dissolving	Melting
1	You start with a liquid and something else, often a solid.	You start with one solid only.
2	The solid mixes into the liquid to make a new liquid.	You turn that one solid into a liquid.
3	You don't need heat.	You have to heat the solid to melt it.

Questions

1 Which of these things is a solution? Sea water, fresh water, Coke, sugar, lemonade, mercury.

2 Why does the water in a swimming pool sometimes hurt your eyes? What is dissolved in the water?

3 Find out what happens when you melt sugar. Ask someone how to make caramel. Your home economics teacher might help.

26. School dinner

Round about the cauldron go; in the poisoned entrails throw.

Double, double, toil and trouble: fire burn and cauldron bubble.

Eye of newt and toe of frog, wool of bat and tongue of dog.

Scale of dragon tooth of wolf...

William Shakespeare wrote this witches' recipe for his play *Macbeth*. Unfortunately he didn't make the witches good scientists. They did not write down a proper description of their experiment. They did not tell us how much of each thing they used. They did not tell us how long they boiled it for. They didn't even say what they were trying to make – but it sounds like Irish stew!

Stew is not a solution. Stew is not a smooth, runny liquid you can see through. The important thing about stew is the lumps floating about in it. A liquid with lumps is called a **suspension**.

Separating salt and pepper

Suppose you had a teaspoonful of mixed salt and pepper. How could you separate them? If you had hours and hours to spare you could pick out all the bits of salt with a pair of tweezers. But it would be more sensible to use a solution.

Salt dissolves in water. Pepper doesn't. Drop the teaspoonful of mixture into a glass of water, and stir. The pieces of pepper will float about in suspension. The salt will dissolve, and make a solution. After a few minutes all the salt will have dissolved. Then you can **filter** the mixture. Pour it through a filter. You can make a filter by pushing a paper tissue into the ring of your finger and thumb.

When you filter a suspension the solid lumps are caught by the filter paper. The liquid goes through. In this case the salty solution will go through. The solid pepper will be left in the paper. To get the pepper back, let the filter paper dry out, and then you can shake the pepper off. To get the salt back is more difficult. You have to get rid of the water by warming it up and letting it evaporate. As the water turns to steam the solid salt is left behind.

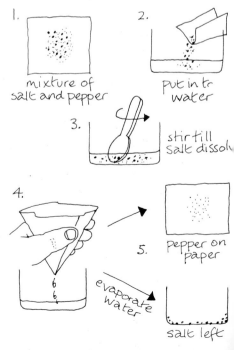

1. mixture of salt and pepper
2. put into water
   stir till salt dissolv[es]
3.
4.
5. pepper on paper
   evaporate water
   salt left

Different types of suspensions

Suspensions need not be solid lumps in liquid. Write down this table in your book.

SUSPENSIONS
liquid in liquid	= EMULSION
liquid in gas	= MIST
solid in gas	= SMOKE
gas in liquid	= FOAM

Life is full of suspensions. A crowded swimming pool is a suspension of bodies in the water. Clouds are suspensions of water drops in the air. Emulsion paint is a suspension of oily drops of paint in water. This is easy to put on the wall and easy to wash off the brush.

What we want 'ere's a nice drop of emulsion!

Questions

1 What is the difference between a solution and a suspension?

2 In your book, make two columns, one for solutions and one for suspensions. Write down each of these things in one of the two columns: salty water, Irish stew, fog, lemonade, milk.

3 Explain clearly how you could separate a teaspoonful of mixed sand and sugar.

4 Copy and complete this crossword.

Across

1 Chlorine does this to the 10 in a 6 (10)

6 You can crawl here, but don't drink! (8,4)

9 This alcoholic solution may be real or light, and it's three quarters of pale (3)

10 The liquid you get by condensing steam (5)

13 You will find these bits of solid in stew, but not in a solution (5)

14 This goes on the wall, and may be a 15 (5)

15 Suspension of liquid in liquid (8)

Down

1 Something that goes into solution has to do this (8)

2 This is what you do to sugar to make it 1 down in 10 . . . (4)

3 . . . but if you try to 1 down the 13 in stew you are bound to do this (4)

4 A suspension of solid in gas (5)

5 You do this to ice to turn it into 10 (4)

7 Lots of things measured in grams rhyme with gases (6)

8 Its eye went into the witches cauldron (4)

11 Water from this doesn't taste much (3)

12 This is what we call 10 when it falls from the sky (4)

27. What's in a vacuum cleaner?

How do birds fly? They flap their wings, but how does this keep them up in the air? Why don't they come crashing down?

Because they push down on the air with each wingbeat, and *the air pushes back.* The force of the wingbeat makes more pressure in the air underneath the wing, and this extra pressure pushes the bird upwards.

Hold your hand in front of your mouth and blow. You can feel the air pressing against your hand. That pressing of the air is **air pressure**.

Air is invisible. You can wave your hand through it with no trouble. Sometimes we make the mistake of thinking that air isn't real. But you can feel its pressure.

push

air pressure increased under the wing

air at higher pressure pushing up on bird and down on ground

Living at the bottom of a sea

We walk about on the surface of the Earth, and above us is the air. It goes up for miles. We live at the bottom of a sea of air. The birds flying about overhead are like fish in the ocean, and we are like crabs, scuttling about on the bottom.

All that air is heavy stuff. We often make the mistake of thinking that air weighs nothing, but it is miles deep, this sea. Your classroom probably contains more than 100 kg of air. On every square centimetre of the Earth it presses with a force of about ten newtons. That means there is a force of about ten newtons pressing on your fingernail.

This force presses in all directions. Not just downwards, but sideways and upwards as well. That is why we are not pushed down flat. All that weight of air does push down from above, but we can still walk about because the air pushes up from below as well.

10 newtons

high pressure

Air inside pushes out equally in all directions

Why are bubbles round?

When you blow a bubble it forms a sphere. Bubbles are always round – never flat, or square. One reason for this is that bubbles have more pressure inside than outside. The

extra pressure inside pushes in all directions. The walls of the bubble are pushed outwards the same amount in all directions. So the bubble is round.

What's in a vacuum cleaner?

A vacuum is nothing. Empty space. No air. A vacuum cleaner has a fan inside which blows some of the air out. This lowers the air pressure inside. You never get a complete vacuum inside. But the lowering of the air pressure makes more air rush in from outside, and the air rushing in scoops up the dust and dirt from the floor.

You can test this yourself. Put the plastic nozzle on the vacuum cleaner. Switch it on. Put your hand over the end. (Use a piece of paper if you like.) Your hand smacks against the end of the tube, and is stuck there. The air pressure inside is lower than the air pressure outside. So the air outside pushes your hand against the end of the tube. You can feel the force.

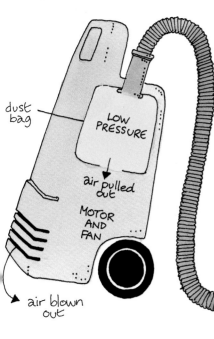

dust bag

LOW PRESSURE

air pulled out

MOTOR AND FAN

air blown out

air rushes in to try and fill up low pressure area inside

Questions

1 Is there a vacuum in a vacuum cleaner? If not, what is there?

2 Hold out one hand in front of you. Air is pushing down, with a force of about ten newtons on each square centimetre. How much is that, roughly, on your whole hand? How is it that you manage to hold it up?

3 Mountain climbers need extra oxygen at high altitudes, because the air pressure is low up there. Suggest reasons why the air pressure is low up mountains. Would you expect blowing bubbles to be easier or more difficult than at sea level? Why?

4 Design and describe an experiment to measure the air pressure inside a vacuum cleaner. You could use a large tin of beans and kitchen scales, and see how much you can reduce the weight of the tin using the vacuum cleaner from above.

Beans

28. Who has seen the wind?

Air pressure is important in our lives because it varies. When the air pressure is different in two places what you get is wind. When the pressure is low inside the vacuum cleaner the wind rushes in. It's almost the same with the weather.

The weather forecasts on television tell us about depressions and highs. What they are talking about is air pressure. Usually we have good weather when there is high pressure. Low pressure seems to bring rain and snow.

Have a look at Map **A**. There is a low-pressure area over Ireland. What does that mean? First of all, there will probably be rain in the middle of the country. Also, air all around that Low will try to rush in and fill it up – just as air rushes in to try and fill up the vacuum cleaner.

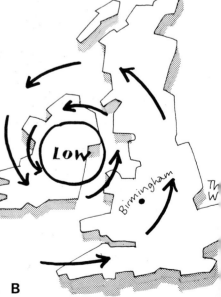

Who has seen the wind?
Neither I nor you.
But when the leaves
          hang trembling,
The wind is passing through.

Who has seen the wind?
Neither you nor I.
But when the trees bow
          down their heads,
The wind is passing by.

Christina Rossetti

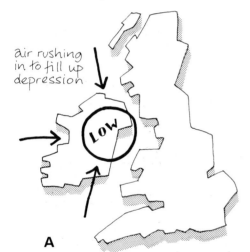

air rushing in to fill up depression

but the earth is spinning

A

But Nature isn't as simple as that. The Earth is spinning. As the air rushes inwards the Earth turns underneath it, and leaves some of the air behind. The result is that the winds blow in a spiral, as shown in Map **B**.

Here is the air trying to fill up the Low. Wind rushes in from all directions, but because of the spinning of the Earth the wind comes in as a spiral.

Now you can see what that depression does for the rest of Britain. The wind is rushing into Devon and Cornwall from the west. Birmingham is getting wind from the south. The west coast of Scotland is getting wind from the south-east.

B

Prevailing wind

There is usually a little wind blowing. On average our wind comes from the west. So we say that in this country our **prevailing winds** are from the west.

In a day or so the prevailing wind will move that depression across the country. It might perhaps reach the North Sea. Meanwhile another one may be coming in from the Atlantic – or with luck some high pressure. Then the weather is likely to be fine.

Hurricane 'DAVID'

The strongest winds

In this country we sometimes have gales. These are winds blowing at up to 45 m.p.h. Storm force winds blow at about 60 m.p.h. The hurricanes that sometimes hit Jamaica have wind speeds of about 75 m.p.h. A hurricane is caused by a huge travelling depression, with its own Low in the centre. The wind howls around it, causing terrible destruction.

Perhaps even more violent are the tornadoes, common in the southern United States. These are much smaller – often only a few metres wide. But the force of the wind can be devastating. One man in Texas was sitting in the bath when the tornado hit the house. The next thing he knew he was still sitting in the bath, but half way up a tree, 300 metres from where his house had been.

Questions

1 In a depression, what sort of pressure is there? Which way does the wind blow? What sort of weather is there likely to be?

2 Suppose the depression shown in Map **A** moved across to the Wash by the next day. Draw a map to show what sort of winds you would expect over the country. Which way would the winds blow in London? In Wales? In Scotland?

3 Look at the weather forecast in the newspaper or on television. Write down in your book a forecast of what you think the weather is going to be like tomorrow. Are there any lows or highs near? Which way are they moving? Will it rain? How strong will the wind be? Where will it blow from? How high will the temperature get? The next day you should note whether you were accurate.

29. Cloud squeezing

What are clouds? From below they sometimes look grey and dull. From above, if you see them from a plane, they look like huge fields of fluffy cotton wool. From inside they look like fog. Fog and mist are just clouds sitting on the ground. They are all made of tiny drops of water. Suspensions of water drops in air.

The steam from a kettle is invisible, but when it reaches the cold air a few centimetres from the spout it cools, and condenses into a mist of tiny water drops, or droplets. This is exactly how clouds are made.

Warm damp air is full of water vapour. The air stays clear as long as the water is all vapour, but if the water condenses into droplets the air goes cloudy.

heat from sun

water evaporates

wind

Where do clouds appear?

Clouds can be formed anywhere when warm wet air is cooled. Air gets cooler as you go higher. So if warm wet air is lifted upwards then it will be cooled, and clouds are likely to appear.

Prevailing winds carry warm wet air across Britain from the west. Here and there are hills, such as the Pennines. When the warm wet air reaches the hills, it has to climb, and it gets cooler, and clouds form. Often there is rain.

Why does it rain?

Clouds are made of millions of tiny droplets of water. Sometimes those droplets come together to make bigger drops, and then they fall as rain.

What is odd is that raindrops grow bigger by not falling. If they fall as soon as they are big enough, then they will still be small when they reach the ground. But if there is wind rushing upwards, and this stops the little drops from falling, they can hang about in the sky. While they hang about they bash into other drops, and get bigger and bigger. So the biggest raindrops have spent lots of time hanging about in the sky.

Getting the wind up

Hot air rises. Air expands when it is heated, and so its density gets smaller. As a result it tends to float upwards through the air all round, just like a cork bounds upwards if you hold it under water and then let go.

When the sun heats a field, the air above the field is warmed, and this warm air will rise, making a **thermal**. A thermal is a rising column of warm air. Glider pilots and birds use thermals to gain height and save wing work.

Hot-air balloons are wrapped thermals. A big bag full of hot air is much lighter than the same volume of cold air. So a hot-air balloon floats upwards into the sky, carrying cargo with it.

Thermals begin over warm fields. If the fields are damp, and the air is damp too, then the thermals carry all that water vapour up with them. That is how the water gets up there.

begins to rise and becomes a thermal

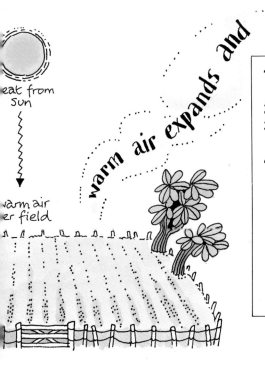

warm air expands and

eat from Sun

warm air er field

Questions

1 What are clouds? What is the difference between dew and clouds?
2 Why do clouds often appear over hills?
3 Explain how warm air can help make raindrops grow bigger.
4 Look at the sky at 8 am, 12 noon, and 4 pm for the next three days. Note how much of the sky is covered (all, $\frac{3}{4}$, $\frac{1}{2}$, $\frac{1}{4}$, or none). Draw a bar chart to show how the cloud cover varies.

   Look in the library to find the names of different clouds (cumulus, cirrus etc.). Draw pictures of them. Note in your record which types of clouds you see. Can you draw any conclusions?

30. Thermals and thunderstorms

The sun heats a wet field all morning. Warm wet air above the field makes a thermal – a column of warm wet air rising into the sky. Rising columns of warm wet air may make thunderstorms.

High in the sky the temperature is low. As the thermal climbs it goes through cooler and cooler air. The water vapour begins to condense into droplets. A cloud begins to form.

The air is still rushing upwards. The droplets are swept up by the climbing air. They collide and run into one another. They combine to make bigger drops. As they get higher and higher they grow bigger and bigger – remember that raindrops get bigger by not falling.

It's cold up there

Strong upward winds, blowing at perhaps 20 m/s, carry these drops many kilometres above the ground. When they get high enough some of them begin to freeze into feathery crystals of snow. Others get so big that they begin to fall in spite of the wind.

The huge drops fall down. The feathery crystals blow up. As they sweep past each other there is 'electrical friction' between them. This is why we get lightning and thunder.

mmm...... Much like the electrical friction between snow crystals and raindrops.

Electrical storms are like stroking the cat. Sometimes when you stroke the cat its fur will crackle under your fingers. This is a result of the electrical friction as your fingers rub over the fur. In much the same way water drops and ice crystals can rub past one another to make lightning.

Across
3 A huge travelling depression may cause this, and bring 75 m.p.h. winds (9)
5 When the air pressure varies, the weather is likely to be this (5)
6 The pattern of wind round a depression, or water round a plughole (6)
7 Flashes caused by electrical friction in thunderclouds (9)
10 A single bit of rain (4)
11 In a cloud, a 10 won't do this until it has time to 8 (4)
12 Feathery white crystal of ice (9)

Down
1 Stand under a thundercloud, and you're not likely to stay this (3)
2 A suspension of tiny water drops in the air near the ground (4)
3 Rain may be caused by warm damp air being blown over these (5)
4 This big beautiful bird of prey uses thermals to gain height (5)
8 In a cloud, a 10 will do this by bashing into other 10s (4)
9 The noise of a wolf, or the wind round a hurricane (4)
11 The air pressure under its wings helps a bird to do this (3)

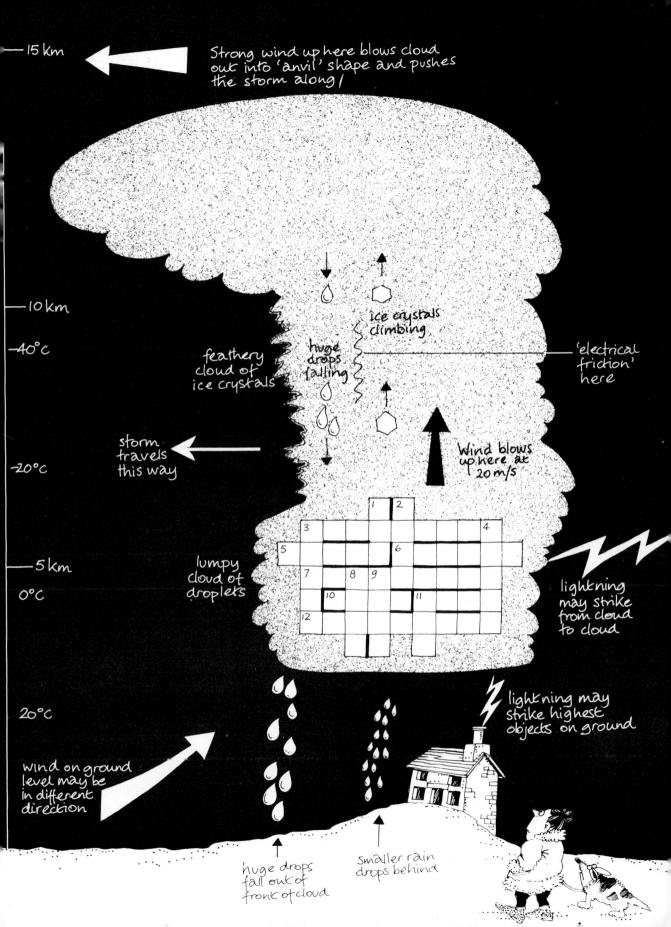

— 15 km

Strong wind up here blows cloud out into 'anvil' shape and pushes the storm along!

— 10 km

— −40°C

ice crystals climbing

feathery cloud of ice crystals

huge drops falling

'electrical friction' here

storm travels this way

Wind blows up here at 20 m/s

— −20°C

lumpy cloud of droplets

— 5 km

0°C

lightning may strike from cloud to cloud

20°C

lightning may strike highest objects on ground

wind on ground level may be in different direction

huge drops fall out of front of cloud

smaller rain drops behind

31. Ice from the sky

Dirty grey snow cloud
Drifting over Derbyshire
Dropping all its cargo
Of slowly falling snow;
Soft flakes, feathery,
Floating through the winter sky
Swirling into drifts so high
That grow, grow, grow.

These pictures are close-up photographs of a snowflake (above) and a slice of hailstone (below). Why do you think they look so different? They fall differently, too.

Angry black thunder cloud
Flashing over Somerset
Rumbling and thundering
And lashing at the ground
Hard stones, hailstones,
Hammering on everyone
Quickly come and quickly gone
The whole year round.

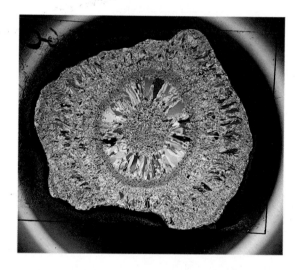

Hail from the violent storms

In a thunderstorm there are violent upward winds. Warm wet air can be blown many kilometres into the sky. Suppose some of the water has condensed into raindrops, and is then flung high up, where the air is well below freezing. Those raindrops may freeze fast.

Fast freezing turns drops into little balls of ice. They just solidify in the shape they are.

raindrop

colder

ice lump

+ rain

wet ice
lump

hailstone

+ raind

wet
hailstone

bigger
hailstone

Suppose some of those round lumps of ice collide with raindrops. The cold raindrops will freeze on top of the lumps. A new shell is formed round the old one.

The water drops freeze so quickly that little air bubbles often get frozen in. A large hailstone may have several layers of ice, with many tiny bubbles of air trapped inside.

Slow freeze for snow

Now forget the storm. It is a cold, winter's day. The sky is a yellowish grey, heavy with water vapour. The temperature is below freezing. The water vapour begins to condense.

Because the temperature is below 0°C the water vapour begins to turn into solid ice, but in this case it happens slowly. A tiny piece of ice appears first. Then more water vapour comes along, and more ice begins to grow on the first piece. It's not dumped on in dollops, as happens in hailstones. This time it grows slowly, to make **crystals**.

Crystals are solids that form slowly in regular shapes. They are often beautiful, as are ice crystals. When they fall from the sky we call them snowflakes.

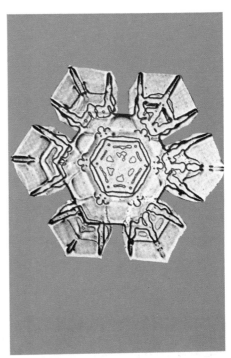

Frost on the ground

During a cold night the ground temperature may fall below freezing. If that happens when the air is damp then the water vapour in the air may condense slowly on to blades of grass, twigs, and fences. In each place ice crystals grow, and we call them frost.

Questions

1 What are the main differences between snowflakes and hailstones?
2 Why don't snowflakes fall from thunderclouds?
3 The basic pattern of a snowflake is hexagonal. Find out how to draw a hexagon. Draw one on cardboard and cut it out. Use the cardboard hexagon to draw round. Then see how many patterns you can make by drawing hexagons touching one another, like snowflakes.
4 Next time there is a hailstorm, try cutting a hailstone in half. Can you see the layers of ice?

32. The skin on the water

There's a funny little insect called the pondskater. It has short front legs, and it runs about on top of the water. The pondskater is not at all heavy. But even so you might expect its feet to sink into the water.

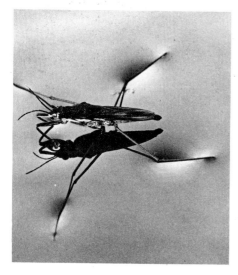

There is also a lizard about 10 cm long which dashes across the water. It's called the Jesus Christ lizard. It runs furiously across, as if it isn't quite sure whether it can reach the other side without sinking.

Why don't they sink?

For the pondskater and the Jesus Christ lizard, water behaves as though it had a skin on the surface. The creatures' legs press a little way into this skin, but don't go through. Instead the skin bends down for them. What you can see is an example of the **surface tension** of the water.

Water always shows this surface tension – as though the surface was pulling inwards, and doesn't want to be broken.

You can see this clearly when you float a paper clip on the surface of water. Take a dry paper clip. Make a sling from a 3 cm square of paper towel or toilet paper. Use the sling to lay the paper clip carefully and gently on the water in a tub or saucer. Then carefully use another paper clip or a pencil to poke down the paper and leave the paper clip floating.

surface
tension

paper clip

The paper clip has a density much greater than water has. Clearly it should sink at once. It floats only because of this curious surface tension. Look at the photograph. You can see how the weight of the paper clip bends the surface of the water. The water surface behaves like a stretched rubber blanket pushing up on the metal.

Smooth water

Have you noticed the smoothness of water when it runs from a tap? It comes out as a smooth round rope, or as smooth round drops. *Always round. Always smooth.* Because the surface tension pulls it *inwards.*

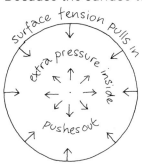

surface tension pulls in

extra pressure inside

Pushes out

Bubbles are also smooth and round. A bubble is a bundle of gas pulled together by a skin of water. The volume of the bubble is fixed by the gas inside. The shape is affected by the skin. The skin of water pulls the bubble into the shape with the smallest surface area, which is a sphere. So the roundness of the bubble shows the tension in the surface.

Put some bubble mixture on a wire loop, blow the bubble half way up, and then stop. What happens? The bubble blows back at you. Try it and see.

The surface tension is always trying to make a bubble smaller. In this case that is possible. The bubble gets smaller by blowing air out in your face.

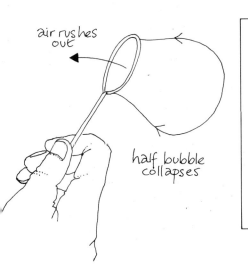

air rushes out

half bubble collapses

Questions

1 Why doesn't the Jesus Christ lizard sink into the water?
2 Why do you think it is called the Jesus Christ lizard?
3 Suggest reasons why a paper towel sling helps when you want to float a paper clip on water.
4 Mr Bubble Mug, a not very well known manufacturer, thinks he has a secret ingredient that will make bubble-gum blow into square bubbles. Write a letter to Mr Mug and explain why you believe he is wrong.

33. The thirsty sponge

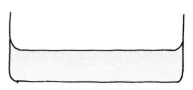

Next time you have a glass of water, take a careful look at the edge of the water, where it meets the glass. The liquid seems to climb a little way up the glass. The glass seems to be pulling the liquid up. Just as surface tension pulls the water, so the glass pulls the water.

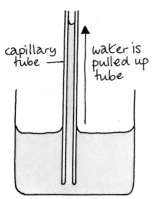

capillary tube

water is pulled up tube

In a narrow glass tube the effect is clearer. The water actually rises up the tube. In a very narrow glass tube, the water almost rushes in. This is called **capillary action**.

A very narrow tube is called a **capillary**. Water is often pulled into and along such tubes. The effect is capillary action, or **capillarity**.

0 hours

Blue flowers
Water evaporates all the time from the leaves of plants. Any flower you pick will dry out and wilt if you don't replace this water. So when you pick a flower you want to keep, you stick the stalk in water.

Put the stalk in a bottle of blue ink, and the petals will turn slowly blue. Why? Because the ink is pulled up into the flower by capillary action, and sooner or later it reaches the petals. This works even better with food colouring.

2 hours

4 hours

6 hours

All living things need lots of water. Most animals have a hole at the top into which they tip water. (Some tip in coffee or beer instead.) But plants don't have mouths. How do they get enough water to live?

Plants have to use capillary action. Their roots and stems have bundles of tiny tubes. Water comes up these tubes from the ground and keeps the plants alive. Even huge trees get their water through the tiny tubes in their trunks.

Soaking up water

Sponges soak up water. Sponges are made up of lots and lots of little tubes with a bit of sponge to hold them together. Dip a sponge in water and the water soaks into the tubes. Take the sponge out and the water stays in. Until you squeeze the sponge, of course.

A paper towel soaks up spilled water, or ink, or milk. A paper towel is full of tiny tubes, although you can't see them as clearly as the tubes in a sponge. The tubes are really just the long narrow gaps between the fibres of the paper, but they work in just the same way.

Keeping the paint on the brush

Artists and painters use capillary action. The gaps between the bristles on a paint brush behave like narrow tubes. They hold the paint in, until the time comes for it to be used.

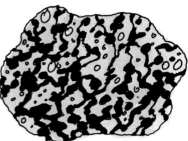

The bristles of a paint brush lie neatly side by side
With narrow grooves between them, into which the paint can slide;
But it doesn't slide right out again – the reason's plain to see:
The artist can rely upon their capillarity.

damp course

Questions

1 What is capillary action?
2 Why does a paper towel soak up water? Why does a Weetabix soak up milk?
3 When they build a house, why do you think the builders put in a damp-proof course between the bricks two or three rows up from the ground?
4 What happens to your clothes when you sit on damp grass? Suggest reasons why this happens. What can you do to stop it?
5 Explain why a candle won't work without a wick (see **24**).

34. Depression in the bath

A **vortex** is a twisting spiral. A tornado is a vortex. You often get a vortex when you pull out your bath plug.

Pull out your bath plug. The water doesn't just run straight down, does it? Often it glugs a bit. Then there is a pause. Usually a little dent appears in the water over the plughole. Slowly, the water begins to turn in a whirlpool round this dent. Suddenly, there is a vortex – a spinning funnel of air down through the water to the plughole itself.

Depression in the bath

When you pull out the plug, that plug hole is the lowest point in the whole bath. It is literally a *depression*. The water above it behaves like the air round a Low in the atmosphere (*see* **28**).

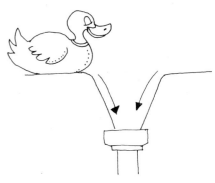

First the water tries to run straight in. But it is slow to start running, held back by what is called **inertia**. Then when it does start running in, it doesn't usually run straight in. Instead it runs in a spiral, like the wind trying to fill up a depression in the atmosphere.

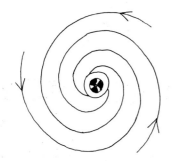

In Britain wind always blows the same way – anticlockwise round a Low. Usually the water does the same. If you look at the next vortex in your bath you may well find it goes anticlockwise.

In Australia the spirals go the other way. Winds blow clockwise round a depression, and bathwater usually runs clockwise in a vortex round the plug hole. That is because if you are in the southern hemisphere the Earth seems to be turning the other way.

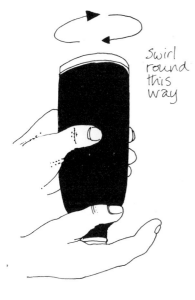

Swirl round this way

What use is a vortex?

There is another reason why the bath water runs in a spiral. The air can get out of the pipe, and will not get in the way of the water. That's why the glugging stops when the vortex starts.

You can use this in a trick to get water out of a bottle quickly. Take an old bottle, such as a large plastic Coke bottle. Fill it with water. Turn it upside down over the sink or outside, and see how long it takes to empty. Time it with a watch.

Now fill it up again. Put your hand over the end while you turn it over. Swirl the whole bottle round violently five or six times. Now take your hand off the top, and see how long the water takes to run out.

air gets in here

When you swirl the bottle you start the water moving round. Then when it starts to run out it forms a vortex in the neck of the bottle. This looks as though it would slow down the flow. Less water can get into the neck.

But it does a vital thing. It makes a path for the air to get into the bottle. If the air doesn't get in the water won't run out. Without swirling the air has to **glug** in. Each glug stops the water dead. The vortex lets the air flow in and the water flow out, both at the same time.

Challenge your friends to a race of bottle emptying. But don't tell them about the secret of the vortex till after you have won!

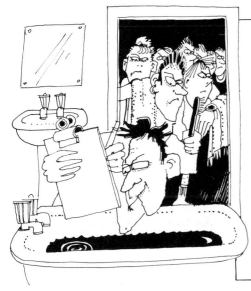

Questions

1 What is a vortex?

2 Which way would you expect your bathwater to spin (a) in France? (b) in New Zealand? (c) at the Mt Kenya Safari Club Hotel, which is exactly on the equator?

3 Conduct a survey of as many of the class as possible. Ask each one to check each bath and basin and sink in the house. Put some water in, leave for several minutes to settle, pull the plug out carefully. Note which way the vortex goes. If possible do each plughole three times. When you add them all together, what percentage spiral anticlockwise? What conclusions do you come to?

35. Daniel Bernouilli

Daniel Bernoulli (say 'Bernooly') was a Swiss scientist. One of the things he worked out was the link between wind speed and air pressure.

Take a strip of paper about 20 cm long and 5 cm or more wide. Fold it into a bridge. Put the bridge on the desk.

Now what do you think will happen if you blow under the bridge? Write down what you think will happen.

Have you written it down? OK, now do it. Blow down under the bridge. What happens? Does the bridge blow away or does it flatten down on to the table? What happens when you blow harder? Why?

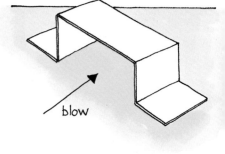
blow

High speed = low pressure

This was what Daniel Bernoulli worked out. The higher the wind speed, the lower the pressure. When you blow under the bridge the air is rushing through the gap. The speed of the air is high. So the pressure is low.

Because the air pressure under the bridge is low, the pressure on top of it will be greater. And so the bridge is pushed down. There is a force pushing down on the roof.

This link between air speed and pressure may not seem important. But it is. There is nothing else to keep aeroplanes up in the sky.

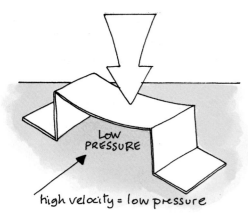
LOW PRESSURE
high velocity = low pressure

This is a slice from the wing of a plane. It is called an aerofoil section. As the plane flies along, the wings cut through the air. Some of the air goes over and some under, as shown in the picture.

Because of the shape of the section, the air that goes over the wing has to travel further than the air that goes underneath.

leading edge

trailing edge

So, if the air is going to meet again at the trailing edge, the air going over the top must move faster than the air going underneath.

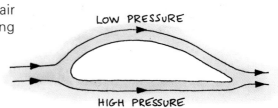

LOW PRESSURE

HIGH PRESSURE

LIFT

But faster flow means lower pressure. So there is lower pressure above the wing than below. Therefore there is an upward force, which is called **lift**.

There is always lift, as long as the wing is moving through the air. This lift is what keeps the plane flying.

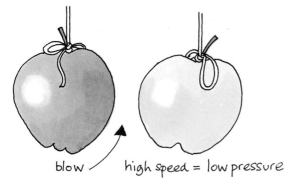

blow   high speed = low pressure

Questions

1 What is the link between air speed and pressure?
2 If you were to hang two apples up on bits of string and blow between them, what do you think would happen, and why?
3 Copy and complete this crossword.

Across

2, 3 down. Brick walls are protected from rising damp by this sort of course (4,5)
5 This pull at the top behaves like a skin on the water, and allows the 9 to run across (7,7)
6 Water is pulled into this very narrow tube (9)
7 The force that keeps planes up in the air (4)
9 Trap on desk confused insect that walks on water (10)
10 The density of an object is its mass divided by this (6)

Down

1 This is what 5 tries to make bubbles (7)
2 You can see this on the water surface under a floating paper clip, or under the feet of a 9 (4)
3 See 2 across
4 Jumbled rats it needs 5 across to keep paint on the brush (6)
5 What a flower has to stand on and drink through (5)
8 This solid form of water is nearly nice (3)
10 up. A twisting spiral you can find in the bath (6)

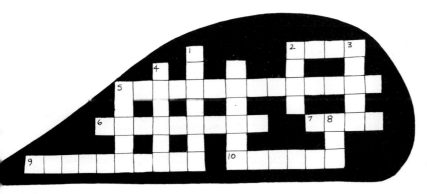

36. Blowing in the wind

If you were a plant, you would want to get your seeds spread out and sown each year. You would need **seed dispersal**. How would you go about it?

Plants can't walk. They can't move about to sow their seeds. Human beings sow lots of seeds – carrots, beans, rice – but only the ones they want. If you were an unpopular plant you couldn't rely on human beings.

Suppose you were a thistle. It would be important for you to get your seeds in the ground. Otherwise there might be no thistles next year. And you know that your seeds aren't going to be spread about by people.

Why bother with dispersal?

If you just drop your seeds by your own stem, they won't do much good. For one thing, there probably won't be room for more than one plant in that spot. So out of all the seeds only one will grow. And if you are still planning to be there next year then none of the new plants can survive.

Besides, what you want to do is expand the tribe. Take over more land. 'Power to the thistle!'

So you have to get your seeds spread out as far as you can. Then with luck each one will fall on a bit of fresh ground which has no thistles yet, and start a new colony.

How to spread your seeds

Plants have found five ways to spread their seeds. The least common is by water. The seeds are dropped in tiny boats, and float away down the river or across the sea.

Some plants have explosive fruit, and fire their seeds as if from a cannon. On a warm autumn day you may be able to hear the popping from gorse bushes.

POP! POP! POP! POP! POP!

Shall we walk or take a dog?

Many plants use public transport – animals. Plants like burdock have hooks round their seeds. They hook themselves on to the fur coats of passing animals, and may be carried for miles.

Blackberries are delicious food for birds. The birds eat the seeds inside, and later the seeds come out in their droppings. Birds also like cherries, but spit out the stones. Either way, the seeds get a ride!

The fifth way is with the wind. If you are still a thistle that is how you would do it. Tie your seeds to a feathery parachute, so that they can be blown away by the wind. Some seeds are so light that they blow away on their own. Others make their own helicopter blades to whirl through the air as they fall.

I'm sycamore competition.

Questions

1 What is meant by seed dispersal?

2 Why do plants need to disperse their seeds?

3 Look in the library to find out about two different plants which disperse their seeds in different ways. Draw pictures of the plants and their seeds and write five lines about how their seed dispersal works.

4 Organize a dispersal competition for the class. Each pair has to make an aircraft from one sheet of paper. Take them all into the playground, and see which flies furthest. Draw a bar chart of the distances they fly, and see what you can work out about the types of design which are best.

37. Rising damp

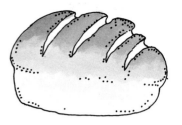

When bakers make traditional bread, they start by kneading the dough. The dough is left in a warm place, and it *rises* – the lump gets bigger. Then it is shaped into loaves and baked. In the oven it rises again. The loaf of bread that comes out is bigger than the piece of dough that went in.

Why should this be? Does everything expand when it is heated? No – it can't be as simple as that. Cabbages don't get bigger when you cook them. Meat usually seems to shrink. Why should bread expand?

What we do know is that gas always expands when you heat it. That is part of the secret. Dough is full of gas. When the gas expands it blows the dough up, like a spongy football.

Yeast is yeast and west is west

Yeast is a tiny plant. It's a boring brown colour, and it has no leaves or roots, because it is a kind of **fungus**. It eats sugar, but in a most peculiar way.

When yeast eats sugar, it doesn't use any oxygen, but it does make carbon dioxide. Because yeast eats sugar in this peculiar way, people have found yeast useful since long before sliced bread was invented.

You can buy dried yeast in packets in the supermarket. If you want to see it at work stir a few grams of sugar into a little warm water, and add a few little bits of dried yeast. The small pellets soften and squidge out. The water goes cloudy. It begins to smell warm and *yeasty*. Tiny bubbles of gas come up to the surface. This gas is the carbon dioxide. It may not look very exciting, but millions of people need that gas for their bread.

You should take yeast, mate — help yer to rise better.

So should you, me dear, then perhaps you'd be better bred.

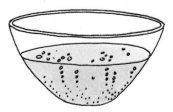

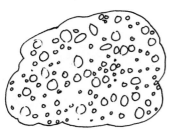

When it gets warm and damp . . .

There are many recipes for bread. Most of these use flour and water, or milk, and yeast. Mix these together, perhaps with some sugar, to make the dough.

Then you have to knead the dough, to make sure the yeast is thoroughly mixed in with the other stuff, and damp, and softening.

Then you leave it in a warm place, and the yeast eats the sugar – or some of the flour. The tiny bubbles of carbon dioxide are trapped by the damp, sticky dough. And so the whole lot is gradually inflated. Blown up like a football. The dough is rising.

Then you put the dough in the hot oven. The yeast is killed by the heat. But the little gas bubbles expand as they get hotter. And so the bread rises again.

And another thing

When yeast eats sugar, it also makes alcohol. Stir yeast into warm grape-juice, and you will finish up with wine. Or start with barley malt, and you can make beer. Bread would be alcoholic, but all the alcohol evaporates in the hot oven.

Questions

1 When you cut a slice of home-made bread, why does it seem to be full of holes? Some breads are flat – chapatis, Pitta bread, unleavened bread. Can you suggest reasons for this?
2 What does yeast eat? What does it make?
3 Find out about fungus in your school library. Write ten lines about it. How many different sorts do you know? Where do they live? How do they spread?
4 Try and get permission to make some bread at home. As soon as you have put it in the oven, write down in your book what you noticed about the dough. How did it feel? How did it smell? How much did it rise? How did you keep it warm? And most important, how did the bread taste?

38. What's in the tube?

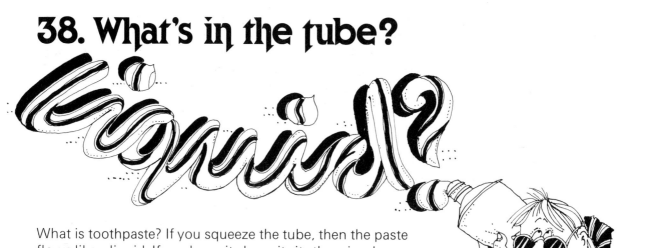

What is toothpaste? If you squeeze the tube, then the paste flows like a liquid. If you leave it alone, it sits there in a lump, like a solid. What do you think it is?

A plastic liquid is a liquid that flows, but only when it is pushed. So toothpaste is a **plastic liquid**. Take the top off the tube, and you can wait all day for the stuff to run out. But if you squeeze the tube anywhere then out it squirts.

When you squeeze the tube you are increasing the pressure inside. Pressure works in all directions, and it doesn't matter where you squeeze the tube; the toothpaste will flow out just the same. But a metal tube is much more likely to suffer metal fatigue and crack, if you squeeze it often in the middle. So to keep the tube happy it's best to squeeze it from the bottom. You can squeeze a plastic tube anywhere!

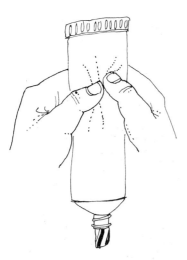

Sticky and springy

Toothpaste is sticky stuff. A bowlful of it would be tough to stir. If you get some on your finger it doesn't run off like water. You have to wash it off.

Toothpaste is also springy. If you hold the tube with the nozzle down you can check this. Squeeze gently until the toothpaste sticks out about 1 cm below the nozzle. Now release the pressure. Usually the toothpaste will spring back a little up the tube.

You can also lay a couple of toothpaste tramlines, say 10 cm long and 2 cm apart. Lay a ruler along them. Now try pushing the end of the ruler gently. First along the line of the rails. Second across the line. Push and then take your fingers away. Does the ruler spring back a bit? It probably won't spring back the whole way, but it may well spring back a bit.

ruler

toothpaste

Try twisting this way

Toothpaste is both sticky and springy. Liquids like this are called **visco-elastic**. *Visco* means sticky; *elastic* means springy. So toothpaste is both plastic and visco-elastic.

The secret of the sauce

A few years ago special new paints appeared, called **thixotropic** paints. They looked like jelly in the tin. Great sticky lumps came out on the brush, and they painted on like slippy liquids. A thixotropic liquid is like a jelly until you disturb it. Then it flows freely. The cans of paint said DO NOT STIR – because if you stir you break the thixotrope, and the paint runs and drips. As a jelly, it won't drip.

Ketchup is thixotropic. That is why it is often hard to get out of the bottle. It sets like a jelly. But with your knowledge of thixotropes you can beat it. First shake the bottle hard before you unscrew the top. The thixotrope is broken. Take off the top and the sauce will run out easily.

Questions

1 Toothpaste is a plastic liquid. Write two more sentences to give examples of things that are visco-elastic, and thixotropic.

2 If toothpaste was as runny as water, it would run off the toothbrush. Suggest reasons why the manufacturers make it plastic.

3 Custard is sometimes thick and gooey. What experiments would you do to find out whether it is plastic, visco-elastic, or thixotropic?

39. Rub-a-dub-dub

Toothpaste is a curious material, packed with scientific interest. Some people also use it to clean their teeth!

Some toothpastes are clear-coloured, jelly-like liquids. Some are white. Some are blue. Some have stripes. All the manufacturers make different claims about how they clean your teeth. But most toothpastes have some chalky grit added. So the paste is a suspension of solid in liquid, and this suspension is important when you clean your teeth.

Why bother to brush your teeth?

When you eat a mouthful of food, some of it will stick to your teeth. Whether it is crisps or custard, some of the food will stay behind, coating your teeth with a thin layer. There is sugar in most of our food; so this thin layer has sugar in it.

Sugar is just the stuff for the bacteria in your mouth which are looking for a place to settle. They camp in a sugary place and feast away. As they eat the sugar they produce acid, which softens the hard shiny enamel on the outside of your teeth. Once the enamel is soft the bacteria can get in and cause decay inside. Then you will need fillings from the dentist, or even worse, you may have to have the tooth pulled out.

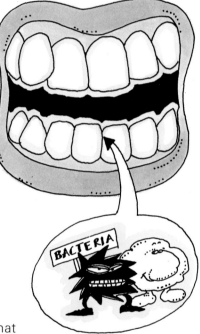

Clean teeth are strong and shiny. If you keep them like that the bacteria will not be able to get through and cause decay. The point of brushing your teeth is to get rid of the thin sugary layer of food.

Just washing your teeth with water is no good. The sugary layer is too well stuck to your teeth. Even the toothbrush may not shift it all. That is where the toothpaste comes in.

Clear gel toothpastes have a special detergent which helps to dissolve the sugary layer, so that it can be washed away. The chalky grit in most toothpastes works rather like polish.

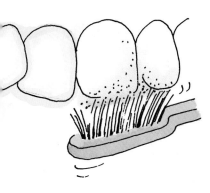

Triumph for abrasion

With your brush you rub the chalky grit against the sugary layer. The grit is harder; so it scrapes the sugary layer away. But your teeth are harder than the chalky grit; so the grit does not damage the enamel of your teeth. This is called **abrasion**. Abrasion means rubbing something away by super-friction.

Sandpaper works by abrasion. If you do woodwork you will use it to scrape away the rough edges and corners of the wood.

At London Airport the landing lights must stay clean. Otherwise planes would not be able to land safely at night. The lights get coated with a layer of oily dust, not easy to wash off. They are cleaned by abrasion. A jet of compressed air loaded with crushed walnut shells does the trick. The crushed walnut shells are just hard enough to scrape off the oily dust, without damaging the glass of the lights.

Questions

1. How does the chalky grit in toothpaste help to clean your teeth?
2. Toothpaste and a toothbrush are excellent for cleaning the lenses in spectacles. Can you suggest reasons why this should work?
3. Make a pencil mark on paper. Then rub it out with a rubber. Watch carefully while you are rubbing. Write down, in terms of abrasion, what happens to the paper, the pencil mark, and the rubber.
4. Parents are always complaining about clothes and shoes getting worn out. Write a short letter to them explaining why this happens.

40. Thirst for blood

Their craving for human blood is never satisfied. By day they are corpses, but at night beware, especially when the moon is full. For then they become the un-dead. They rise from their graves, and suck the blood of their sleeping victims. For they are **vampires**, and people they have bitten turn into vampires themselves.

Vampires have never really existed, but this legend was made popular by Bram Stoker in his book *Dracula*. Count Dracula was a vampire. His face 'was hard, and cruel, and sensual, and his big white teeth, that looked all the whiter because his lips were so red, were pointed like an animal's.'

He cast no shadow in the moonlight. He never ate anything. He could not be seen in the mirror. And he could turn himself into a wolf, or into a bat. But above all, his 'peculiarly sharp white teeth . . . protruded over the lips.'

Once they caught him in the act. 'His eyes flamed red with devilish passion; the great nostrils of the white aquiline nose opened wide and quivered at the edges; and the white sharp teeth, behind the full lips of the blood-dripping mouth, champed together like those of a wild beast.'

In the end the heroes managed to kill the count, and 'in the drawing of a breath, the whole body crumbled into dust and passed from our sight.' But by then he had bitten several victims, including Lucy; 'Just over the external jugular vein there were two punctures . . .' 'Her teeth seemed longer and sharper than they had been in the morning. In particular, by some trick of the light, the canine teeth looked longer and sharper than the rest.'

She died, and became un-dead, and at night went round biting children in the neck, and sucking their blood. So the heroes had to lay her to rest, with a heavy hammer and 'a round wooden stake, some two and a half or three inches thick and about three feet long.' Her fiancé 'placed the point over the heart, and as I looked I could see its dint in the white flesh'. Then he struck with all his might.

'The Thing in the coffin writhed; and a hideous, blood-curdling screech came from the opened red lips. The body shook and quivered and twisted in wild contortions; the sharp white teeth champed together till the lips were cut and the mouth was smeared with crimson foam.'

Questions

1. What was the size of the stake in centimetres? What was its volume in cm^3? (1 inch is about 2·5 cm.)
2. Which are your *canine* teeth? Draw a diagram of your top front teeth to show them. Are they sharper than the others? Do they hang over your bottom lip? Suggest reasons why they are called canine.
3. Where is your jugular vein? Can you feel a pulse in it? Would the neck be a good place for a vampire to bite? Why?
4. Vampires seem to survive on a diet of human blood only. Do you think this is possible? Write a short letter to your Home Economics teacher. Explain the problem. Say why you think it might or might not be possible. Ask for sensible advice.
5. The book is not entirely clear, but it seems that the vampires suck the blood through their hollow canine teeth. Scientists in Transylvania say that the holes are not more than 2 mm in diameter. Suppose that you could live on one litre of milk a day, and nothing else. Describe an experiment to find out how long you would take to suck up 1 l of milk through two straws. If the diameter of a straw is 3 mm, what is its area of cross-section? Guess how much more slowly the blood would come through holes only 2 mm across. (*Hint*: what is their area of cross-section?)
6. Real vampire bats live in south America. They live on blood, but they don't suck it from their victims. Find out in the library how they get the blood, and who their victims are.
7. One thing that is supposed to keep vampires away is garlic. What is garlic? Look it up in the library and write ten lines about how it grows, what it looks like, and what else it is useful for.
8. Dracula may have been able to turn into a bat. But is it possible that he could fly, with his 'cloak spreading out around him like great wings'? Could you fly, with wings stretched between your arms and legs? DO NOT TRY IT. Instead, suggest some simple experiments to test whether your arms could push down with enough force to keep you up. With your arms spread out, how much weight can you lift?
9. When you suck liquid through a straw, or a tooth, you are making use of a difference in pressure. Explain carefully how you make this pressure difference. What does it do? How does air pressure help you? Using a long tube of clear plastic, how high can you suck water up? Draw a bar chart of the power of suction of the class. Who, in your class, would make the best vampire?

41. Lung power

Did you hear about the time when Mr Universe failed? He was standing in a television studio with very large muscles and very small swimming trunks. He was invited to sing his favourite song while lifting 150 kg.

He could sing the song (more or less). He could lift the weights. But he could not do the two things at the same time. He started the song, but had to stop dead when he lifted the weights.

Can you suggest any reasons why he could not lift and sing together?

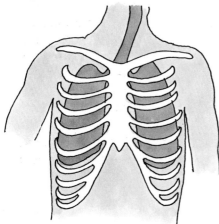

Inspiration!

Your lungs are like a couple of balloons inside the front of your ribs, near the top. Count down two or three ribs from your collarbone, half way to your shoulder. That's where your lungs are. They are a bit bigger than your fists, and pink and spongy. Pink because they are full of blood. (If a person smokes they may be brown and slimy.) Like a sponge because they have to have a big surface area, so that the air which comes in can meet all the blood inside the lungs.

The stuff your lungs are made of is like very thin skin, so thin that the air can get through.

To start with you must get the air in. First, use your muscles to lift your chest. Expand your rib cage. At the same time you lower your **diaphragm** (say 'diafram') – though you can't see that.

What this does is lower the pressure inside your chest cavity, around your lungs. Because you have made the space inside bigger the pressure inside is reduced.

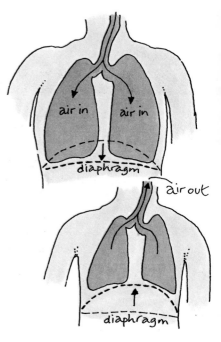

The pressure of the outside air is now higher than the pressure in your lungs. So air rushes in (as long as you have either your mouth or your nose open) and fills your lungs. You have breathed in. This is called **inspiration**.

Now let your chest go down again. Lift your diaphragm. The pressure inside goes up. The air rushes out. You have breathed out.

The trap-door

There is a small snag. You have only one mouth. This has to cope both with air for your lungs and with food for your stomach. You won't come to much harm if you get air in your stomach, though you may burp a bit. But you must not get food into your lungs. You would choke.

To stop you breathing food into your windpipe you have a trap-door for safety. This is a flap of skin called the epiglottis. When you eat, it folds over the top of your windpipe to keep the food out.

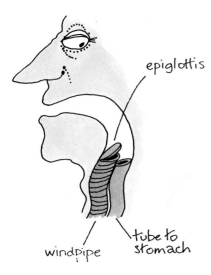

epiglottis

tube to stomach

windpipe

The Valsalva manoeuvre

But if you want to make a great effort you shut your epiglottis anyway. When you want to push very hard on something, or lift a big weight, you often grunt. What you are doing is filling your lungs with air and then shutting the epiglottis. Then you can use the high pressure inside to make your chest hard and firm. Then it's much easier to push or lift. This is called the Valsalva manoeuvre.

But doing it makes you grunt. When your epiglottis is shut you may be able to lift a heavy weight, but you won't be able to sing.

grunt

Questions

1 Why did Mr Universe fail?
2 What pushes the air into your lungs when you breathe in?
3 What is the name of the flap of skin that folds over your windpipe? What is it for?
4 Think of a way in which you might be able to measure the volume of your lungs. You might use a balloon, or take a deep breath and blow bubbles under water into a measuring jug.
5 Some tennis players grunt. At what point during the game do they grunt, and why?

42. Don't hold your breath!

The world record for staying under water and holding the breath is more than 10 minutes.

Don't try and beat this record. It could be dangerous. But see if you can hold your breath for **one** minute. Take one deep breath, shut your mouth, pinch your nose, and sit still for one minute.

Can you do it? Some will find it easy, but for others it is quite difficult.

The easy way

There is an easy method. First, take ten slow deep breaths. Each time breathe in as much air as you can. Wait two seconds. Breathe out slowly, and as much as you can. Really empty those lungs. Collapse the chest. Make it tiny.

Wait two seconds. Then breathe in again.

After ten deep breaths like this, take one even bigger breath. Try to hold it. One minute is easy. Why is it easy? Because all the deep breathing helps delay a message from your brain.

Auto or manual?

Most of the muscles in your body are under your control. You decide when to stick your tongue out. You choose when to bend your wrist. But some muscles you can't control. Your heart is a muscle. You can't stop your heart beating. Your lungs aren't muscles. Your lungs are controlled by the muscles that work your chest and the muscle that is your diaphragm. You can breathe in when you want. You can breathe out when you want.

But most of the time you don't have to think about breathing. You don't have to say 'Time to breathe in . . . Time to breathe out . . .'. That is all automatic. It happens all the time, even when you are asleep. That is because your brain decides when you should breathe, even if you aren't conscious of it.

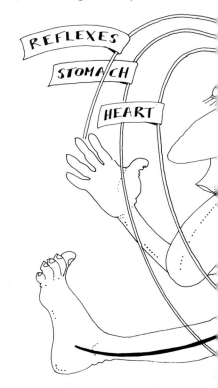
REFLEXES
STOMACH
HEART

How not to suffocate

Your brain needs oxygen. Without oxygen it will die, and so will you. Oxygen is brought to the brain from the lungs by your blood.

Now suppose the brain starts getting a bit less oxygen than usual. It sends out urgent messages to the chest and diaphragm, saying 'Breathe in, you fools.'

When you hold your breath for a minute, the brain will send these messages, and you will be forced to take a breath. That is why holding your breath is difficult.

But what if you take those deep breaths first? What they do is to sweep all the rubbish out of your lungs. Usually your lungs are half full of old air which just stays there, along with some carbon dioxide. Deep breathing empties your lungs properly, and then lets them fill with fresh air. So when you start holding your breath your lungs contain more oxygen than usual. And so your brain takes longer before it starts to worry about the oxygen shortage.

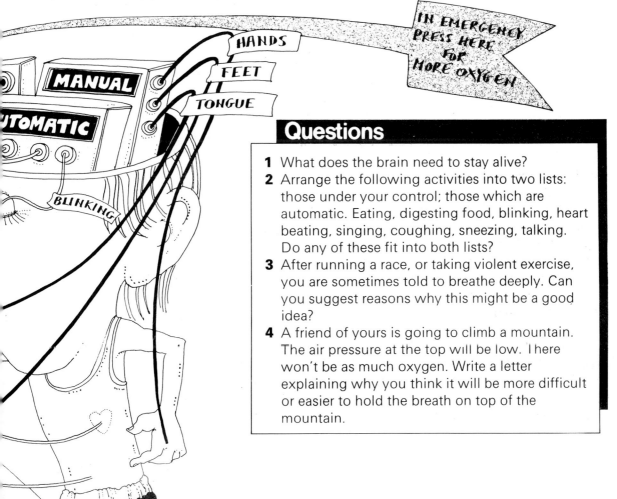

Questions

1 What does the brain need to stay alive?
2 Arrange the following activities into two lists: those under your control; those which are automatic. Eating, digesting food, blinking, heart beating, singing, coughing, sneezing, talking. Do any of these fit into both lists?
3 After running a race, or taking violent exercise, you are sometimes told to breathe deeply. Can you suggest reasons why this might be a good idea?
4 A friend of yours is going to climb a mountain. The air pressure at the top will be low. There won't be as much oxygen. Write a letter explaining why you think it will be more difficult or easier to hold the breath on top of the mountain.

43. Breathing under water

If you want to breathe under water you need gills, a snorkel, or bottled oxygen. Let's have a look at all three.

All animals need oxygen; carried in the blood, it gets pumped round the body. It allows the food to be used, and so provides energy. Without oxygen, the brain packs up. That is why people die of drowning or suffocation.

What's so great about being a fish?

Fish have no lungs. They can't breathe air in, as we can. But they still need oxygen in their blood. They get the oxygen directly from the water.

The water in seas and rivers has air dissolved in it. Just as lemonade has carbon dioxide in it. You can't see the air dissolved in water. But when you *start* to heat a kettle you can hear it coming out. The first faint hiss, before it starts to sing, is the air coming out of the water.

Instead of lungs, a fish has **gills**. These are like bony fans at the back of its mouth. As the fish swims it lets water stream in through its mouth. The water runs over the gills, and out through the gill openings just behind them.

The gills take the oxygen from the water, and put it into the fish's blood. They also take the carbon dioxide from the fish's blood and let that out into the water. So the effect of the gills is to get rid of the carbon dioxide from the blood, and put in fresh oxygen – just like our lungs.

Our lungs can't cope with water. If we want to go under water we have to find ways of getting oxygen gas to our lungs. The easiest way is to use a snorkel.

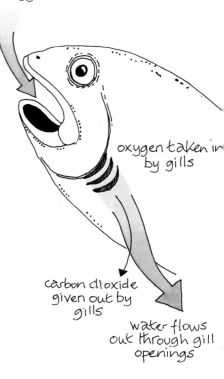

water flows in through mouth with oxygen dissolved in it

oxygen taken in by gills

carbon dioxide given out by gills

water flows out through gill openings

Keep the tube short

With a bent tube called a snorkel you can swim along, looking down, and still breathe air. That means you can take your time to watch the fish, or look for treasure. You don't have to keep coming up for air.

When you dive down, your snorkel fills up with water. Then you must remember not to breathe in! Next time you come up, blow out the water, and you're ready to breathe again.

This is just like the whales. Whales aren't fish, but mammals. They have lungs. They have to come up to the surface for air. When they breathe out, the water vapour in their damp breath condenses.

If you see a whale spout, it's really only breathing out.

What you can't do is try and go deeper by using a longer tube. Each time you breathed out it would fill up with carbon dioxide. And this is what you would breathe in next breath.

But there is a worse problem. You could never breathe in. Two metres down the water pressure is so high you could not expand your chest. We are used to just the pressure of the atmosphere. Two metres under water the pressure on the outside of your chest is 20 per cent higher. You would find breathing in almost impossible.

That is why you must use bottled air if you want to swim deep. Most divers have tanks of compressed air on their backs. When you take it down with you, and it's under pressure, then you can breathe easily. For deep diving you need special mixtures of gases, and expert knowledge.

Questions

1 What do fish use to breathe under water?
2 Where do fish get their oxygen from?
3 A few years ago, snorkels had a bend at the top, and a table-tennis ball in a cage. Can you suggest reasons for this? What would happen if you dived down with this snorkel? Is that a good idea? Suggest reasons why snorkels aren't made like that today.
4 There is a water sports festival at your local pool. A stunt man plans to sit on the bottom in the deep end and breathe through a garden hose. Write a note to him to try and explain why this might not be a good idea.

44. Blowing hot and cold

Once upon a time the winter was so long and so cold that a Centaur came down out of the woods looking for food. He was half man and half horse and had never been into town before.

Don't worry, he's armless!

At the third cave he smelled something tasty. He stuck his head in and was amazed by what he saw. A woman was sitting on a rock blowing on her hands.

Why do you blow on your hands?

To keep them warm.

The Centaur wondered how this could be. Did she have a furnace inside her? Just then her breakfast came and the woman started blowing on her porridge.

Why do you blow on your porridge?

To cool it down of course.

The Centaur was so astonished that he tried to scratch his head. This was a mistake....

He laid himself out with his hoof.

You don't need porridge . . .

Try it yourself. You don't need porridge.

First, hold the palm of your hand 5 cm in front of your mouth. Open wide. Breathe out a long, slow pant. Warm, isn't it? And damp.

Now hold your hand 10 cm in front of your mouth. Blow a sharp blast, as if you were whistling. (No need to make a noise.) This time it feels cold. And dry.

Water needs heat to evaporate

A cup of tea cools down mainly because water evaporates from the surface. This takes heat out of the rest of the tea. Water needs heat to turn into steam.

Water evaporates from surface. This takes heat from tea and so cools tea down.

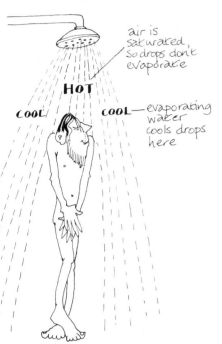

air is saturated, so drops don't evaporate

HOT

COOL — COOL — evaporating water cools drops here

The palm of your hand is slightly damp with sweat. When you blow hard across it you help to evaporate some of the dampness on your skin. Evaporating the water also takes heat away, and so your hand feels cool.

But when you pant slowly straight on to your hand, your breath, straight from your mouth, is warmer and damper than your hand. So the water vapour in your breath begins to condense on your hand. Your hand feels a little damper. And when the water condenses it gives up the heat which was needed to evaporate it. The condensing water gives heat to your hand.

If you have a hot shower, you may notice that the water is cooler on the outside than in the middle of the shower. This is because the drops on the outside evaporate and cool. The ones in the middle don't evaporate much. There is so much water in the middle that the air is full of water vapour. The air is **saturated**. No more water will evaporate, and so the drops stay warm.

Questions

1 What does water need to make it evaporate?
2 Find out from books in the library where Centaurs were supposed to live and when. What did they do?
3 Write 100 words for the Centaur News explaining why you can cool porridge by blowing.
4 Suggest reasons why dogs pant.
5 Lots of heat is needed to evaporate water. You might expect that countries surrounded by sea would stay cool. You might also guess that places that are a long way from any sea might get very hot – or very cold. Find out whether this is true. Where are the hottest and coldest places on Earth?

45. Personal explosions

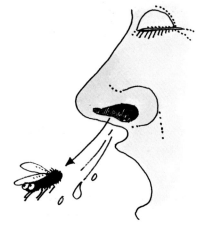

There are three kinds of personal explosions. Sneezes and hiccups are **involuntary**. That means that you can't sneeze or hiccup when you want. Right now you can't count '1 . . . 2 . . . 3 . . .' and then hiccup. And usually you can't stop when you want, either.

Coughs are partly **voluntary**. If you want, you can cough right now. And even when your throat tickles a lot, you can usually almost stop coughing. Just by deciding to stop.

Aaatishoo

Inside your nose is skin that is very sensitive. As you breathe air in through your nose the big chunks of dust are caught by the hairs in your nostrils. You must not get dirt in your lungs.

But tiny bits of dust can get through. Then they may irritate the skin. The skin reacts against the irritation. You take a deep breath and explode it out through your nose to get rid of the dust. That's a sneeze. You have little control. The sneeze happens automatically.

Sneezing is meant to be a self-defence system to guard the lungs. But sometimes it is set off by pollen. Pollen is the yellow 'dust' made by grasses to fertilize their flowers. In the early summer the air can be thick with it on hot dry days. People who are sensitive may sneeze all day. This is called **hay fever**.

When you have a cold, the extra fluid in your nose may start you sneezing. The trouble with this is that your sneezes may fire the germs out at more than 100 m.p.h. (40 m/s). The germs may zoom several metres through the air, and infect everyone else in the room.

Hic

Hiccups are caused by eating too fast. Or you may have irritated your stomach with coke, or something else it doesn't like. What happens is that your diaphragm starts twitching.

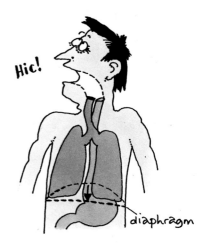

Each time your diaphragm jerks down it yanks air in at the top. At the same time the top of your windpipe shuts, just as if you put a cork in. The 'Hic' can be uncomfortable in your throat.

Next time you are in the bath, or have a deep basin full of water, you can make artificial hiccups. Pull out the plug. Let the water run out for a few seconds, till it's running fast. Then jam the plug back in. You will get a *thunk* like a hiccup.

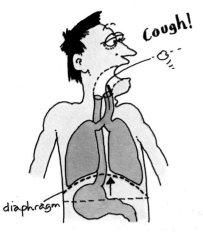

Er-hem

Clearing your throat is a good way to get rid of a little bit of irritating stuff. But if you want more power then you have to cough.

Sometimes you get a crumb in your windpipe. It 'goes down the wrong way.' Your breathing tubes have hairs in, like your nose. They react violently to crumbs. You take in a deep breath. You close the top of your windpipe. You build up the pressure inside. Then you open up and blast out the invaders. Less violent coughs are useful to clear your throat. You usually have some control over them.

Questions

1 What causes hay fever?
2 Why do we say that sneezes are involuntary?
3 Find out in your library the world records for sneezing and hiccupping.
4 Suggest reasons why people usually shut their eyes when they sneeze.

46. The naked ape

There are 193 different kinds of apes and monkeys, and 192 of them are covered in hair. The only ape that is not hairy all over is Man. Most mammals are covered with fur – cats, dogs, rabbits – but not human beings.

The biggest bit of fur you have is on top of your head, where you probably have more than 100 000 hairs. Each hair grows about 1 cm each month. As many as 100 fall out every day. Cutting your hair won't make it grow any faster, but it will get rid of the untidy split ends.

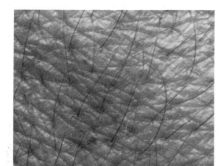

Follicles and roots

Look at the skin on the back of your hand. Use a magnifying glass if possible. The hairs come out of little dents, called **follicles**. Each hair grows from a root under the skin. As it grows out through the follicle the hair brings a little grease with it. This is why your hair can get greasy.

Long, greasy hair gets tangled, and gathers dust. This looks a mess. When you wash and comb and brush your hair it looks better. But it isn't healthier. All hair is dead; so nothing can make it healthier. After washing and brushing the hairs lie smoothly side by side. They reflect the light, and so your hair shines.

Pull out one of your hairs and have a look at it. The root is the long lump on the end. Compare it with your friend's hair. (DO NOT pull out your friend's hair, unless you are invited to do so.) Which is longer? Thicker? Curlier?

Why aren't we covered with fur?

When a baby chimpanzee is born it has a thick head of fur, but the rest of its body is almost naked. So people are a bit like chimps that have never grown up. Chimps need their fur to keep warm. What use is our bare skin?

We have sweat glands on our bare skin. When we get hot, one way we cool down is by sweating. Millions of years ago this may have been important. When early humans moved away from the trees they had to start running after their food. Perhaps then they needed to keep cool during the day more than they needed to keep warm at night.

Head lice

Your hair is like a jungle – a great home for wildlife. Any one of us can catch head lice, and one child in every fifteen does. You can discourage lice by brushing or combing your hair every night, but if your hair touches the hair of someone with lice they can hop across from head to head.

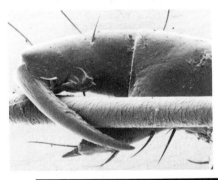

◁ A head louse gripping a human hair

Your head may itch. You may find the eggs or the 'nits' in your hair. Nits are the empty eggshells of lice that have hatched. If you think you may have head lice, ask someone to look – a nurse, or one of your parents. Having lice is not dirty or disgusting. They often settle in the cleanest hair. But you must get rid of them. You will need the right lotion, and the right help. The whole family should be treated to make sure the lice are wiped out.

Split-end magnified × 320
▽

Questions

1 How are humans different from other apes?
2 What are the differences between your hair (the one you pulled out) and your friend's hair? Draw pictures of them.
3 Make a survey of the hair in your class. Beside each name note length, curliness, thickness, and colour. Make pie charts to display the information you gather.
4 Some shampoo manufacturers claim that protein shampoo makes hair healthier. Write a letter to the Customer Relations Department at Sham-Poo plc, explaining why this is impossible.

47. The cat's whiskers

catstache

moustache

Have you ever thought about why cats have whiskers? All cats have these funny tufts of long hairs. They have one or two on each cheek, a few above each eye, and a bunch on each side of the upper lip.

The softest touch

Many people think that your fingertips are the most sensitive bit of your body. They think you can feel the softest touch with your fingertips. It's not true. Reach up with your hand. With your fingertips, very gently touch the hair on top of your head.

Where do you feel the touch first? In your fingertips, or on your head? Write down where you feel it first.

You can try this with a friend. Both keep your eyes shut. Touch the other's hair as gently as you possibly can. Each say the moment you first feel contact.

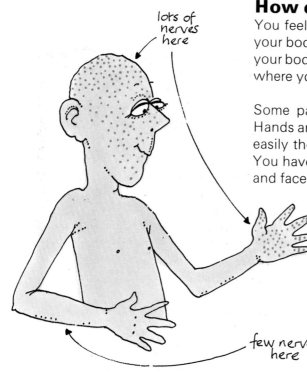

lots of nerves here

few nerves here

How do you feel?

You feel touch with your nerves. You have nerves all over your body, but most of all in your skin. The skin is the bit of your body that's nearest to the outside world. So your skin is where you need to feel any contact.

Some parts of your skin are more sensitive than others. Hands and face are sensitive. You can feel any contact most easily there. This is because you have lots of nerves there. You have more nerves per square centimetre in your hands and face than you have in your arms or legs.

Move quickly!

Help, it's hot!

What a nerve!

Nerves are the electrical alarm system of the body. They send messages to the brain to tell it about what is happening. If you stick your finger in a flame, one set of nerves rush a message to your brain to say 'Help! Things are getting hot down here.' Another set brings a message back to the muscles in your arm: 'Get out of there. Fast.'

But how can we feel with our hair? Hair is dead. Hair has no nerves. Hair can't send messages to the brain, even if it is nearby.

The skin underneath has nerves. When your hair is touched it pushes on your scalp. That is how you feel the touch. Your scalp – the skin on your head – is very sensitive.

The cat's whiskers

Cats are **carnivorous predators**. This means they eat meat, and they hunt. At least they are supposed to hunt. In real life most cats lie around sleeping, and wait for someone to open a tin of catfood.

But if a cat does go hunting, then its whiskers will help it to feel its way past obstacles in the dark. No cat will catch many mice if it keeps crashing into table legs and tripping over trainers. The whiskers are the cat's distant early warning system.

tickle

Questions

1 Suggest reasons why it hurts to have your hair pulled, but not to have it cut.
2 Why do you think cats have whiskers in front of their ears?
3 Can you suggest a simple experiment to find out whether you have more nerves in your tongue, or on the inside of your cheeks? Do the experiment. Write down your conclusions.
4 Tickling is gentle unexpected touching. Why is it so effective to tickle the back of someone else's neck? Why can't you tickle yourself? If you use a feather, or a piece of paper, can you tickle someone else better? Or yourself?

48. Look out! The world is upside-down

You can't really look out at all. What you can do is open your eyes and let the light come in. Your eyelid is like a bath plug. Pull out the plug and the water pours into the plug hole. Pull your eyelid out of the way and the light pours into your eye.

Green light comes from the grass. The grass looks green. Blue light comes from the sky. The sky looks blue. Your eyes use the light to make a picture of the world, and tell your brain what is happening outside.

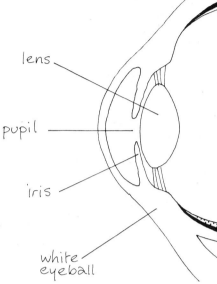

lens

pupil

iris

white eyeball

Eyeballs

An eyeball is like a big squishy marble. Shut your eye. Touch it gently with your fingers. You can feel that it is round and soft. But don't press too hard; that hurts. NEVER put anything in your eye.

Look carefully at your friend's eye. The eyeball is white round the outside. You may be able to see the thin red lines of the blood vessels that bring blood to the eye.

The coloured part is called the **iris**. What colour are your partner's eyes? A few people have eyes of different colours – perhaps one blue and one brown. The job of the iris is to control how much light gets into the eye through the hole in the middle.

The black hole

The dark bit in the middle of the iris is called the **pupil**. This is just a hole, like a plug hole. The light from outside gets into your eye through this hole, the pupil.

Your eye works like a camera. At the back of a camera is the film. At the back of your eye you have a **retina**. Here the light makes an image. The retina collects the light and sends messages to your brain, with a picture of the world you see.

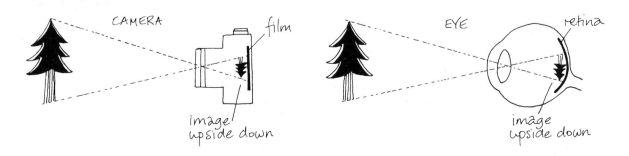

CAMERA   film

image upside down

EYE   retina

image upside down

Why is the world upside-down?

The image or picture on a camera film is upside-down. This doesn't matter. You can turn the photograph over. But the image on your retina is also upside-down. You can prove it with a piece of paper and a sharp pencil.

Poke three or four holes in the paper, close together. Hold the paper about 10 cm from your eye, and between your eye and the window, or a bright light. Look at the holes you made.

Hold the pencil upright, point upwards. Move the point carefully up in front of your eye. The point should be 1 or 2 cm in front of your eye. BE VERY CAREFUL NOT TO POKE YOUR EYE. The pencil may just touch your eyelashes.

retina

The light from each hole streams past the pencil and on to your retina. There it makes a patch of light, with a shadow of the pencil point. These shadows, like the pencil, are point upwards. But when you see the shadows you see the pencil point coming *down* into the holes. You see the shadows upside-down.

In fact you see the whole world upside down, just like the camera. But your brain has learned to turn it all over again. As a baby you learned which way up things really are. So your brain turns everything over, and now you see the shadows the wrong way up!

Questions

1 What does your pupil do?
2 Suggest an experiment that will let you find out what happens to your partner's pupils (a) in bright light (b) in dim light.
   Do the experiment. Write down your results.
   What is the iris doing?
3 What shapes are the pupils of cats, dogs, and other animals in bright light?
4 Copy and complete the wordsquare.

1 The coloured bit of your eye
2 Picture on the retina
3 Your pupils get small in this light
4 Where the light collects at the back of your eye.

49. Superbrain

Do you ever forget things? Archimedes forgot everything when he jumped out of the bath. We are all a bit like that. We get absent-minded. Would you like to remember everything? To remember really well, what you need are tricks.

A **mnemonic** (say 'ne-monic') is a trick to help you remember something. Professional memory artists use mnemonics all the time. Here is a simple one. The colours of the rainbow are Red, Orange, Yellow, Green, Blue, Indigo, and Violet. The first letters of those words are ROYGBIV. Remember the word Roygbiv, and you can remember the colours of the rainbow. Or remember '**R**ichard **O**f **Y**ork **G**ained **B**attles **I**n **V**ain'.

How to stick memories together

No one knows exactly how the brain works. We do know that we remember things best in groups. You can recall your three times table partly because it is a group of numbers. A good group makes a mnemonic.

Learn this list. Read it through four times. Say it aloud if you can. Better still if the whole class can chant it together.

1 is a BUN

2 is a SHOE

3 is a TREE

4 is a DOOR

5 is a HIVE

6 is STICKS

7 is HEAVEN

8 is a GATE

9 is a MINE

10 is a PEN

You must know this list off by heart. Practise till no one can catch you out. When you know it you will have a mnemonic you can use to amaze everyone. You will remember the shopping list. You will be able to recite the top ten in the charts. You can use your one-is-a-bun list to remember any other list of ten things. What you must do is group the two lists in pairs. Suppose there are these ten things:

1 Cow 2 Cabbage 3 Ladder 4 Pencil 5 Fire-engine
6 Cup 7 Bicycle 8 Cat 9 Fish fingers 10 Teacher

When you see this new list, don't try and remember that 1 is a cow. You know that **1 is a bun**. So connect the cow and the bun. Make up a silly picture in your head of a cow and a bun together. A cow eating a bun . . . or a cow shut in a giant bun, like a mooing hamburger . . . or a cow standing up on top of a huge sticky bun. The sillier the picture, the better.

As soon as you have made up a picture, go on to the next. **2 is a shoe**. How about a cabbage stuffed into a shoe . . . or someone trying to clean a pair of shoes with a cabbage?

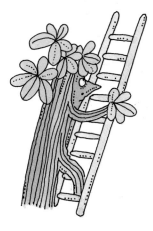

3 is a tree. A ladder up a tree is easy – but you will remember a silly picture better. How about a tree climbing a ladder?

You will need about 20 seconds for each word. When you get to the end you will find you can easily remember the whole list. You can recite it backwards. You can recall number 5, or any other number. Everyone will think you have a superbrain!

MNEMOSYNE

Questions

1 What is a mnemonic? Why is it useful?
2 Ask your partner to make you a shopping list of ten things, and read them to you slowly. You do the same. Can you both remember everything?
3 Who was Mnemosyne (say 'ne-mossiny')? Look her up in an encyclopaedia or under Greek myths in the library.
4 'Thirty days hath September, April, June, and November . . .' Can you complete this mnemonic? What is it useful for? How many other mnemonics can you find? And how many can you remember?

50. Double-glazed coffee

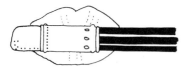

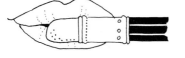

Take a wooden pencil with a rubber on the end.

Hold the painted wood against your top lip. How does it feel? Cool? Warm after a few seconds?

Now hold the metal against your top lip. Does it feel warmer or colder?

And last, the rubber. How does that feel, compared to the metal and the wood?

How can this be? The pencil is all one piece; if one end was hot the heat would flow along to the other end. It must all be at the same temperature. Yet the various parts feel quite different on your skin.

heat conducted
into metal

heat conducted
away by metal

warm
lips

metal band

To understand what is going on you have to think not about temperature but about heat (*see* 19). The pencil is a bit cooler than you are. When you first touch the wood, heat begins to flow from your lip into the pencil. This is *conduction*. The pencil feels cool.

Within a second or two the heat has warmed up the bit of the pencil touching your lip. Wood is a bad conductor of heat; so the heat doesn't go any further. So the pencil suddenly feels warm. Because where it's touching you, it is as warm as you are.

Metals are good conductors
But the metal band is different. When you touch the metal, it quickly conducts the heat away from your lip. Metals are good conductors. Heat flows easily through all metals. Because the heat is flowing out, your lip feels cold.

The rubber is a bad conductor. It quickly warms against your lip, and never feels cool at all.

Insulation with air is the thing
To make winter feel warm
as the spring
So invest in a layer
Of lovely warm air
Trapped in holes tied together
with string.

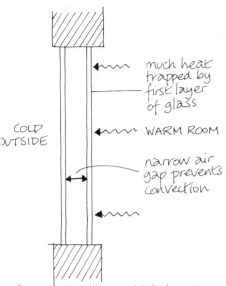

much heat trapped by first layer of glass

WARM ROOM

narrow air gap prevents convection

COLD OUTSIDE

Conduction low, so little heat reaches outer layer of glass. Heat trapped by double glazing.

Bad conductors

Bad conductors of heat include cork, wood, fabrics – especially wool – and air. Air is good at convecting heat round a room. But it conducts it badly.

That is one reason why wool is so warm – both for the sheep and when you wear a sweater. Wool traps pockets of air between its crinkly fibres. The air can't circulate to convect. It conducts badly. So inside the wool you feel lovely and warm.

The reason for double glazing windows is to make a barrier of air to keep the heat inside the house. This saves on heating.

The worst conductor of all is space – a vacuum. That is why the vacuum or thermos flask is so good at keeping coffee hot. The glass conducts heat badly. The vacuum doesn't conduct heat at all. There is nothing to convect between the walls. And they are silvered, to cut down on radiation. Remember how heat radiates from the fire (*see* **19**). Silver reflects this radiant heat back to the hot coffee.

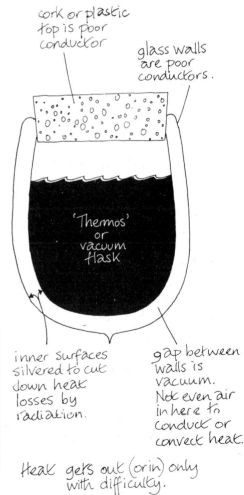

cork or plastic top is poor conductor

glass walls are poor conductors.

'Thermos' or vacuum flask

inner surfaces silvered to cut down heat losses by radiation.

gap between walls is vacuum. Not even air in here to conduct or convect heat.

Heat gets out (or in) only with difficulty.

Questions

1 Why does a rubber feel warm if you touch it?
2 Here is a list of floor materials: concrete, wood, clay tiles, cork, lino, carpet. Arrange them in order of heat conduction, with the worst conductors first.
  Which would you like to have in your bathroom, and why?
3 Your younger brother describes a string vest as 'a lot of holes tied together with string'. How would you try to explain to him how it might keep him warm?
4 You are marketing manager for Keep'ot Thermos Flasks. Write the script for a tv advertisement (90 words or less) to explain why your flasks will keep coffee hot even in freezing cold weather.

51. Blue hands

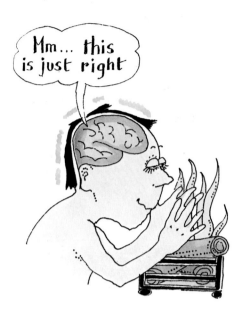

Mm... this is just right

We think of blue hands as being cold. We say *blue with cold*. What we mean is that when someone's face and hands get very cold all the blood runs away from the skin. So the skin goes pale, and can look blue, in contrast to its usual pinky colour.

That change of colour is not just for show. The human body is a cunning machine. There are good reasons for blue hands.

First things first

The bit of the body that matters most is the brain. If the brain dies the rest has had it! So the body is geared to the defence of the brain.

Now the brain is rather delicate. It needs to have lots of blood, with lots of oxygen. And it likes that blood at 37°C. The rest of the body sometimes has to suffer, while the brain stays warm.

When you go out in the snow, or jump into icy water, the body reacts with alarm calls sent to the brain: 'Help; it's cold here.' To which the brain replies 'Tough. But make sure I keep warm.' And what happens is that the corners of the body are allowed to get very cold, so that the brain can keep warm. They are, if necessary, shut off from the usual heat services.

You may be cold, but keep me warm.

Heat is carried round your body by your blood. So if your hands are cold from making snowballs, the blood running through your fingers will get cooler and cooler. This cool blood is unpopular with the brain. So the **capillaries** – tiny blood vessels – in your hands are narrowed down. Less blood runs through them to get chilled. Your brain is kept warm. But your hands go white, and bluish. If they are not rescued in a few hours you will get frostbite.

The same goes for your face. Where it sticks out into the cold air and snow your skin will go white. The brain is protecting itself by closing down the blood vessels in your skin.

Stay cool

When you get hot, it's just the opposite. The brain doesn't want to overheat. So all the blood vessels in your skin are opened wide. Lots of blood runs through. You go pink, or even red in the face.

You sweat; water comes through your skin, and evaporates from the outside. This takes heat away from the skin. (Remember the surprised Centaur?) All these are the brain's tricks for getting extra heat out of the body.

People have been ingenious and clever in designing fridges, anoraks, and thermos flasks. But their own bodies are much more skilful at controlling the flow of heat. The brain is brilliant at heat control.

A house with central heating usually has a **thermostat**. A thermostat is a machine that keeps the temperature constant. It switches the heating on if the temperature drops, and switches it off if the temperature rises. That is what the brain is – a brilliant thermostat.

Questions

1 Why do your hands go blue when you make snowballs?
2 Suggest reasons why athletes have red faces when they finish their events.
3 Suppose you went for a walk in Canada during the winter, when the temperature is below about −20°C. You would begin to feel a prickling in your nose. Each time you breathed out, the water vapour in your breath would freeze, and little icicles would appear in your nose. Suggest reasons why the warmth of your nose does not stop this happening.

52. Shivering fluff

Most people have temperatures around 37°C. So do most birds. So do bats, and rats, and alley cats. What do they have in common? They are all **warm-blooded.**

This means that all these creatures keep their body temperature constant. They have different ways of doing it, but they all keep warm.

Birds fluff their feathers up when the weather is cold. On a frosty morning even the smallest sparrows look big. They fluff all their feathers up so that a layer of warm air is trapped round their skin. This helps to keep them warm. It's just like wearing a string vest.

Most animals have a thick layer of fur to trap the air. We don't have fur but we still try to fluff it up in cold weather. What we do show in the cold is goose pimples, or goose bumps. This is our pathetic attempt to fluff up our fur. The hair we have is so thin and useless that all we get is lumps on the skin.

A more effective human trick is shivering. If you get very cold you will start to shiver. Your muscles twitch rapidly and uncontrollably. You shake all over.

This is another example of the cunning brain at work. Your muscles are not very efficient. When they work they generate heat. That's why you get hot when you run. If your muscles shiver all at once they will generate heat, and this will warm you up. You don't have to run – just stand and shake.

What about the fuel bills?

Being warm-blooded is good, because we have been able to develop our brains. That has been possible only because we have been able to build a constant-temperature body to attach to them.

But there is a price to pay. Food. Keeping warm is so important that quite a lot of the body spends all its time looking after the temperature. Three quarters of the food we eat is used just to keep us warm. Out of every four mouthfuls you take, three are used only for central heating.

That's an expensive way of keeping warm. All warm-blooded animals have to eat an enormous amount of food, just to stay warm-blooded.

Questions

1 Why do birds look big in cold weather?
2 Sometimes your teeth chatter when you get very cold. Suggest reasons for this.
3 We go to a lot of trouble to keep ourselves warm in winter. Make a list of five things we do to keep our bodies warm, and another list of five things we do to stop the rooms in our buildings from getting cold.
4 Copy and complete this warm-blooded wordsquare:
Write two sentences to explain why the thing down the side of this wordsquare is good for keeping you warm, even if it is full of holes.

1 Your muscles make heat when you do this in the cold
2 This is used for keeping the 10 constant
3 A rotten conductor of heat, and not slippery
4 Snowflakes are built up from crystals of this
5 What there is in a 7
6 When you get cold you may get these pimples
7 Empty space – between the walls of a thermos flask
8 What warm water does to become vapour
9 Another name for water vapour.
10 . . . and so it brings this down.

53. Cold blood

Get up early in the morning in Texas and you will see an amazing sight. Hundreds of men going to work in cowboy hats. But 250 million years ago there would have been something much more interesting. As the sun rose the ridges of ground would be packed with dimetrodons. Every one would be facing north or south. There might even have been a stegosaurus or two. They also would be standing sideways to the rising sun.

We think that these dinosaurs were **cold-blooded**. Cold-blooded animals do not heat their own bodies. They stay at the same temperature as their surroundings. They warm up in the sun, and they cool down at night.

What's good about cold blood?

The advantage of being cold-blooded is you don't have to eat much. Warm-blooded animals use three quarters of their food just to keep warm. So cold-blooded creatures can eat much less. A rattlesnake is quite happy if it gets a good meal once a month.

But the cold of the night slows these creatures down. The dinosaurs could hardly move in the morning, until they were warmed by the sun.

The heat we get from the sun is radiant heat (remember **19**). This warms every square centimetre it hits. So for maximum heating the dimetrodons turned their sides to the sun. The fins on their backs were giant heat collectors. Just as you would hold out the palms of your hands to warm at the fire, so dimetrodons used to hold out their fins to warm in the sun.

Cold-blooded animals like the hotter countries of the world. You don't find nearly as many in cold places. In the zoo you will notice that the reptile house is always kept warm. All reptiles are cold-blooded, and their warmth has to come from outside.

If you keep a crocodile in your bath its temperature will stay the same as the bathwater.

The good life – in Death Valley

Death Valley is America's most famous desert. Seven thousand square kilometres of sand, salt, and rocks. Almost no trees or bushes, almost no water, and almost no people. The telephone directory for the entire area has just seven small pages.

The sun shines all day, every day. Even in mid-winter the temperature zooms up each day to 30°C in the shade. People and other warm-blooded animals don't like this sort of place. They sweat too much and dry out. There isn't enough water, and there isn't any food to eat.

For thousands of rattlesnakes, however, Death Valley is paradise. They don't sweat. All reptiles have waterproof skins. They can't dry out in the sun, and they need very little water. Also, they need very little food.

Questions

1 Suggest reasons why the crocodile would stay the same temperature as the bathwater.
2 What use to the dimetrodon was the fin on its back?
3 From the library, find out the names of five warm-blooded animals and five cold-blooded animals. What are the most common cold-blooded animals in this country?
4 A friend of yours has just been given a pet rattlesnake. Write a letter and explain why the rattlesnake will need to be kept warm, and why it will not need as much food as the dog.

54. Survival of the fittest

The animals and plants in Death Valley survive by being unusual. There is a little bush called the desert holly. It looks like ordinary holly, but the leaves are white. This bush pulls up salty water from the ground. When the water evaporates from its leaves, crystals of salt are left behind. The white salt crystals reflect the fierce heat of the sun away from the leaves. So the plant survives.

Fitting the conditions

Nature is cunning. She knows that conditions vary from place to place and from time to time. So she provides a variety of plants and animals. Some of them must be able to survive.

Here in Britain there is a moth that is usually speckled white, but every generation has a few black ones. In the country the black ones don't last long, because they show up easily on light-coloured tree bark. The birds eat them all.

But in industrial areas it's the other way round. The black moths survive. The smoke turns the tree bark black, and the white moths stand out like sore thumbs and get eaten. The black moths are hard to see. They survive.

The black smoke does not turn moths black. The black smoke changes the conditions. With black bark the black moths are **fitter** for the place they live in. This is one simple example of **survival of the fittest**.

The *fittest* does not mean those who run fastest. It means those who best fit into their place in the world.

Variation

Two hundred years ago no one could have guessed that the black moths might take over Birmingham. No one can now guess what will make **you** fittest for the life ahead.

Look around the class. All the children are different. Some have curly hair. Some have blue eyes. Some have brown skin. Some can run fast. Some can scream loudly.

At school the best things to be good at are probably exams, whispering, and not annoying teachers. The colour of your hair, your eyes, and your skin doesn't matter much.

What does matter is that there is a huge variety of people. Apart from identical twins we are all different. We all look slightly different; we all have slightly different skills. Any one of them could come in useful. Some pop singers make lots of money by screaming!

Questions

1 Which animal is the most likely to survive in harsh conditions?
2 Would you expect desert holly to grow in this country? Suggest reasons.
3 What would happen to the moths and why, if all industry in Birmingham stopped?
4 About one person in ten can roll up the tongue to make a tube. Do a class survey to find out how many there are. Suppose there is a world-wide flood of soup. We have to survive by swimming with our hands and sucking up the soup through a layer of liquid mustard. Who would be able to do it? Who would survive? Who would be the fittest?

55. Whatever happened to the dinosaurs?

A hundred million years ago there were dinosaurs all over the place. Huge plant-eaters like brachiosaurus, more than 10 m high at the shoulder. Big meat-eaters like tyrannosaurus, who was as tall as a giraffe and had teeth up to 20 cm long. Where did they go?

There were crocodiles about, too. Crocodiles are almost the same now as they were a hundred million years ago.

Adaptation

As time goes by, conditions change. If the animals are going to survive, then sometimes they need to change too. We have seen how moths can effectively become black. In much the same way other animals must adapt to a changing environment.

Way back in the past, there probably used to be giraffes with short necks. As more and more animals wanted to eat the same trees, so the giraffes with the longer necks got better off. They became the fittest. In the end, only the long-necked giraffes survived. They had adapted to the new conditions. **Adaptation** means changing to fit the new conditions.

Living conditions change over thousands of years. The animals which adapt are fitter than those which cann adapt.

Extinction

Seventy million years ago there was a big change. Living conditions became difficult. We don't know exactly what went wrong, but the dinosaurs began to disappear. Soon they had died out altogether. They were **extinct**.

The problem may simply have been the weather. When we have a cold winter we can turn up the central heating. We can put on more jumpers. We can buy thermal underwear. And we are warm-blooded. That means that we can keep our bodies warm by eating.

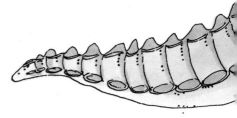

The dinosaurs seem to have been cold-blooded reptiles. When the winters got worse and worse they became slower and slower. Remember, reptiles need the heat from the sun to stay warm and lively. So as the weather got colder the dinosaurs would have found it more and more difficult to collect their food.

The crocodiles may have slipped into the water to keep warm. But the reptiles on land were in deep snow and deep trouble. The dinosaurs could not adapt to the changing climate. They were not fit for these new colder conditions, and they did not survive.

Questions

1 Meat-eaters are called **carnivores**. Which dinosaur was a carnivore with 20 cm teeth?
2 Plant-eaters are called **herbivores**. Find out the names of three herbivorous dinosaurs. Draw pictures of them.
3 Suggest reasons why the dinosaurs became extinct.
4 Copy and complete this crossword.

Across
5 Desert in California alive with rattlesnakes and other reptiles (5,6)
6 Nature's way of choosing which animals live on – those which best fit their environment (8,2,3,7)
11 A huge carnivorous dinosaur, with 20 cm teeth (13,3)
13 Landscape too dry and empty for warm-blooded animals (6)

Down
1 This type of animal needs warmth from the sun to get it going in the morning (4-7)
2 The fittest of these herbivores had long long necks (8)
3 Moths in these areas are mostly darker than they are in the country (10)
4 How many million years since the extinction of the dinosaurs? Three score and ten (7)
7 This variety of looks and skills is how Nature makes sure that there is a fittest (9)
8 What your skin won't be if you are cold enough to have goose pimples (4)
9 An animal that eats grass (9)
10 Stegosauruses used to live in this lone star state . . . (5)
12 . . . warmed by this sort of heat from the sun (7)

56. Pins and needles

Sometimes, when I was a kid, I got a splinter in my finger. Usually my mother used a needle to dig it out. This hurt a bit, but often the splinter hurt more if it was left in.

Before she stuck the needle in my finger she always put the needle in a match flame. I asked why, and she said the flame was to **sterilize** the needle, and kill any germs on it. Otherwise the germs would get into my skin.

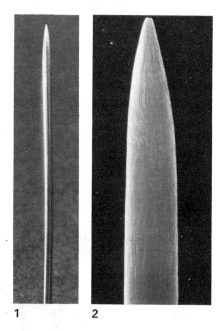

I was never quite sure whether she was right. I knew that germs could get into food, but why should they be on a needle? Needles are made of steel, shiny and sharp and clean. But I believe her now, because I have seen these photographs.

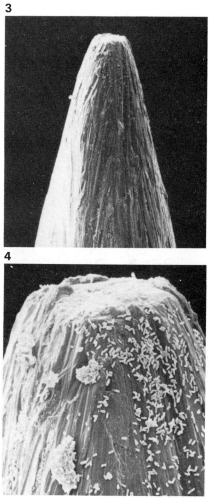

1 The point of a pin, magnified 5 times. This is what you can see if you look at a pin with a magnifying glass. The point looks sharp, but not quite completely smooth.

2 At 30 times magnification the pin point doesn't look nearly so sharp. There seems to be a flat tip instead of a point. The steel looks less shiny and clean.

3 Pin point, magnified 120 times. The end is completely flat – not sharp at all. Down the sides are deep grooves, where the steel has been gouged away to make the point.

4 Giant close-up. 600 times magnification reveals an army camped on the flat end of the pin. An army of bacteria. These are the germs my mother warned me about.

Prick your finger with the pin in these pictures, and some of the bacteria will get off the pin and into your skin. If they get into your skin they may make another camp there, and start chewing up your finger. This is called **infection**. Your finger will get sore, and may even go septic.

How to prevent infection

Don't stick pins or needles into your fingers – or into your friends. If you have to get a splinter out, use a needle. Needles are sharper than pins. Hold the point in the flame of a match or a candle for five seconds. That will kill the bacteria.

If you cut or graze your skin, wash the wound clean with warm water and a little soap. Get help for any bad cut. What the doctor or nurse often does is put some antiseptic on the wound, to kill any germs there. Then they cover the wound with a dressing, to keep other germs out.

Washing your hands helps to get rid of some of the germs on them. Wash before meals and you will not take so many germs into your mouth with your food.

Questions

1 What lives on the point of a pin?

2 Suppose you are seized with an overwhelming desire to stick a pin into your friend. Why is this not a good idea? What should you do to sterilize the pin? How could you test in a laboratory to find out if the pin was clean?

3 Write down at least one sentence to show the meaning of each of these words: *bacteria, germs, infection, sterilize.* You may find the library useful.

4 Vampire bats are dangerous because they carry disease. Write a letter to Count Dracula (Castle Dracula, Transylvania). Suggest politely that his habit of biting people (*see* **40**) might be harmful to their health. Explain why regular brushing with good toothpaste will help keep his teeth in better shape, and also help to prevent the spread of disease.

57. A shot in the arm

Some time this year you may have to have an injection. Many people worry about having injections. This is silly. They don't really hurt at all. You may find it better if you look the other way when the needle is going in.

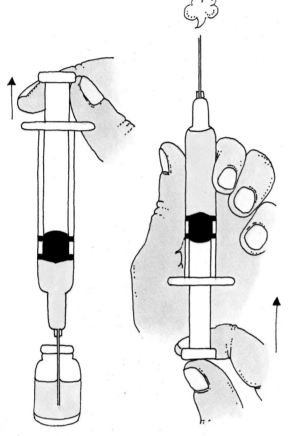

What's an injection?

Injecting means *pushing liquid into.* Doctors and nurse usually use syringes. They put the needle into the liquid, pu the plunger up a bit to suck some of the liquid into th syringe, turn it point up and push the plunger a little to ge the air out. Then they put the needle through your skin an press the plunger to push the liquid into your arm.

They always take care to use sterilized needles. Otherwis you might get a bunch of unwelcome bacteria. Usually the use a new needle for every injection. The new needles ar sterile in their packets.

An injection is a short-cut into the bloodstream. Man poisonous snakes use the same method. They bite you wit their long teeth, and inject you with poison along th grooves in the teeth – like Dracula in reverse.

Why do we need injections?

Most of us have injections to **immunize** us – to stop us from getting a disease. When you get immunized, you won't get medicine in the syringe. What you will get is germs. The point of the injection is to give you the disease.

Two hundred years ago cows often used to get sores, or 'pocks', on their teats. Milkmaids sometimes caught the disease from the cows, and got the sores on their hands. A country doctor, Edward Jenner, noticed this. But, more important, he noticed that these milkmaids didn't catch smallpox. At that time hundreds of people died of smallpox, but not the milkmaids who had had cowpox.

On 14 May 1796 Jenner deliberately gave cowpox to 8-year-old James Phipps. He made two scratches on the boy's arm, and put in stuff from a cowpox sore on the hand of milkmaid Sarah Nelmes. Then, a few months later, he tried to give him smallpox. James Phipps did not catch smallpox. He had become **immune**.

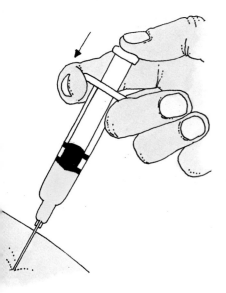

You don't catch things twice

Your body has its own defences, and builds up armies of defenders. If you catch measles your body builds an army of measles defenders, and you are very unlikely to catch measles again. You have become immune to measles.

What Jenner realized was that if you catch a mild form of the disease you get protection from the worse form. Cowpox and smallpox are closely related. Infection with the mild cowpox made the milkmaids immune to the deadly smallpox.

And that is why we have injections today. By getting small doses of germs we build up immunity to killer diseases. Smallpox has been wiped out by immunization.

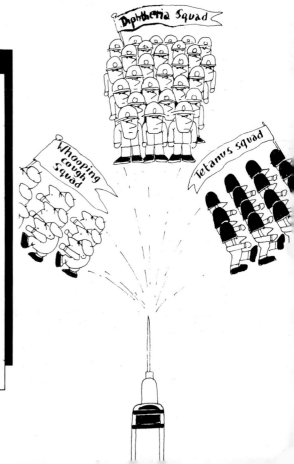

Questions

1 Write down what you understand by the word *immune*.
2 Suggest reasons why milkmaids did not often catch smallpox.
3 Explain carefully, in terms of pressure and density, how a syringe is used to put liquid from a bottle into your arm. (*Hint*: look at the diagrams).
4 Write a letter to a younger brother or sister, explaining why it might be a good thing to have an injection.
5 Find out what immunization you have had. You probably had the 'triple' injection against diphtheria, whooping cough, and tetanus, when you were a baby. What else have you had? Why might you need more if you visit another country?

58. The life that lives on you

The population of Planet Earth is nearly five thousand million people. On your skin there are more than five thousand million animals and plants. They live there. Your skin is their world.

Skin is amazing stuff. It's waterproof, and keeps in all your insides. It's flexible, and bends when you do. It's tough, and not easily torn or cut. When it does get damaged it repairs itself. And all the time it flakes away at the surface, getting rid of dirt and rubbish. Meanwhile fresh new skin grows from underneath.

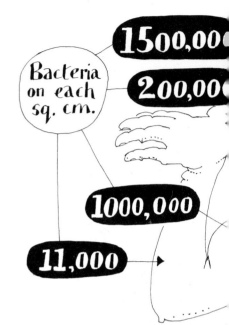

Bacteria on each sq. cm.

1500,000

200,000

1000,000

11,000

hair

pore

HUMAN SKIN

hair follicle

epidermis

dermis

Why aren't you ill all the time?
Not just your skin, but the whole world is full of bacteria. They are minute plants, so small that even magnified a thousand times they would only be as big as pinheads. Many diseases are caused by bacteria, but you don't often get ill. Why? For two reasons.

First: your skin is thick and tough, so bacteria can't get through. If nasty ones settle, they are soon swept off on the flakes of dead skin.

Second: your skin is covered in *good* bacteria. They don't do any harm, and they seem to keep the bad guys away. All those millions of bacteria on your skin are probably protecting you from disease.

One in four Europeans died
The human flea breathes through holes in its sides, and has its heart in its back. It can jump about 20 cm high and 150 times its own length. The dreaded bubonic plague (say 'playg') is spread by its cousin the rat flea. The bacteria that caused the plague were carried by the rat fleas, and injected every time they bit an animal or a person.

There isn't much plague about now, but it hasn't disappeared. Human fleas are no longer common in this country, but many people still get bitten by cat fleas and dog fleas.

Demodex

You mite be surprised

Just above your eyes there's a creature much smaller than a flea. It is a mite, and its name is Demodex. Long and thin, it lives in the follicles of your eyelashes. Luckily, it doesn't seem to do much harm. Your best defence against Demodex, and against the bacteria that causes acne, is plenty of soap and water.

All these things that live on us are called parasites. Dogs have parasites. Cats have parasites. Even parasites have parasites;

Big fleas have little fleas upon their backs to bite 'em
And little fleas have smaller ones, and so ad infinitum.

Questions

1 What is a parasite?
2 You are often told to wash your hands, after you have been to the toilet, and before you eat. Suggest reasons why this washing might be useful.
3 What can you find out in the school library about bubonic plague?
4 Your small brother has fallen over and grazed his knee. He complains loudly when your father washes it, puts on antiseptic, and covers it with sticking plaster. Write to him explaining how important it is to stop huge numbers of outside bacteria from getting through the skin.

59. Rapunzel, Rapunzel, let down your hair

Rapunzel was a German girl, 'the most beautiful child under the sun'. That was what the Grimm brothers said when they wrote a story about her. But when she was twelve a wicked witch took her away from her home.

The wicked witch locked her up in a tower. It had no stairs or doors, and only a little window high up in the wall. Being locked up made Rapunzel sad. She had no science lessons or P.E.

Two years later the King's son came by. As he rode through the forest he heard her singing from the window. He thought she sounded great, but he could not get into the tower.

After a bit of detective work he found out how to get up to her room. He stood under the window and shouted 'Rapunzel, Rapunzel, let down your hair'. She had lovely long hair, 'as fine as spun gold'. As soon as she heard him call she untied the plaits. She twisted the hair round a hook by the window. The hair then fell 20 ells downwards. The handsome prince climbed up . . .

Tr la la...

The prince was kind. Rapunzel laid her hand in his, and said 'I will gladly go with you'. They had some terrible adventures, but in the end they lived happily ever after.

Questions

1 She could not jump down. The prince could not climb up, even from his horse. Guess the minimum height of the window above the ground.

2 In the library, find out what was the longest hair in the world. Should Rapunzel have been in the Guinness Book of Records?

3 How long was an ell?

4 Would her hair have been strong enough for the Prince to climb? Describe an experiment to measure how much weight one of your hairs will carry. This is called the **tensile strength** of the hair (*see* **3**). Look in **46** for an idea of the number of hairs on your head. What would be the tensile strength of all your hair together?

5 Rapunzel was blonde. Would you expect her hair to be stronger or weaker than yours? How would you allow for this in your experiments?

6 Describe an experiment to find out what force is needed to pull one hair out by the roots. USE YOUR OWN HAIR ONLY. Is this more or less than the tensile strength? Why does this matter?

7 Why did Rapunzel wind her hair round a hook by the window? (*See* **4**.) How was friction helping her to meet the Prince?

8 When you have worked out the questions above, do you think the Prince would have been able to climb up? Would you have advised him to take off his armour at the bottom?

9 Surely Rapunzel could not have had head lice? (*See* **46**.) Just to be safe, how could the Prince have avoided catching them?

10 Write a letter to the wicked witch. You suspect Rapunzel has been visited in her room. You don't like the Prince, because he was a bully at school. Suggest to the witch several different ways of keeping the Prince out. She should use scientific methods. She does not understand science. You must explain carefully how each method will work. Give her the benefit of your 'scientific eye'.

Jumbo crossword

Across
1 Solid, liquid, and gas (5,6,2,6)
11 Italian money, confused liar (4)
12 In this world divers feel weightless but need oxygen tanks (5-3)
13 A thousand of what follows – e.g. grams (4)
14 Claps mixed up the skin on top of your head (5)
16 This valley is alive with 32 (5)
18 Hail and snow are both made of this solid form of water (3)
19 A one-litre cube has sides this many cm long (3)
20 Low friction lets people do this on snow (3)
21 Refuse – short for number (2)
22 Acceleration = ____ of change of velocity (4)
23 This book – reading it should help you to look at the world with one (10,3)
26 Worn round the circumference of necks (4)
27 You will get a depression in the bath when you pull this out (4)
28 Rapunzel's hair was 20 times longer than this (3)
30 Number of dimensions in a flat surface (3)
32 Noisy, venomous reptiles of the desert (12)
34 Your feet leave these in dewy grass (6)
37 Where Archimedes liked to soak and think (8,6,5)

Down
1 Rising columns of warm wet air may produce these lightning generators (13)
2 The energy you get from a fire, and dimetrodons got from the sun (7,4)
3 Your weight comes from the pull of its mass on yours (5)
4 Containers which keep heat in or out – often used to keep tea hot (7,6)
5 Idles mixed – but you won't do this if there's too much friction (5)
6 Liquid fuel or lubricant (3)
7 Because the brain keeps warm, this is a danger for frozen hands and feet (9)
8 Mother, Newton worked it out, F = __ (2)
9 Onion vapour brings these to the eyes (5)
10 Warm buildings for cold-blooded in zoos (7,6)
14 Shadows show the time on this type of clock (7)
15 Water pulled up walls by capillary action (6,4)
17 Frog's contribution to witches' suspension, or the last part of Archimedes to leave 37 (3)
24 Carrier of electrical messages in your body is never scrambled (5)
25 What the dinosaurs became, because they could not adapt (7)
29 Bit of face, next to mouth, with lots of 24s (3)
31 A thousand in a litre (2)
33 You need 50 kJ of heat to boil the water for a cup of this (3)
35 There will be friction when two surfaces do this together (3)
36 3.14159...

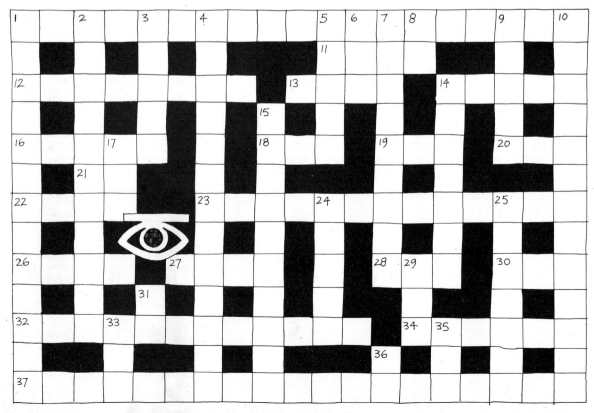